Nylon

The Manmade Fashion Revolution

USE
ME!

Nylon

The Manmade Fashion Revolution

A Celebration of Design from Art Silk to Nylon and Thinking Fibres

Susannah Handley

BLOOMSBURY

To the cherished memory of my husband Roger

First published in Great Britain in 1999 by
Bloomsbury Publishing plc
38 Soho Square
London W1V 5DF

Commissioning Editor Sarah Polden
Designer Satpaul Bhamra
Additional Picture Research Laura Fidment
Index Peter Barber

A CIP catalogue for this book is available from the British Library

ISBN 0 7475 3445 4

10 9 8 7 6 5 4 3 2 1

Colour separation by Tenon & Polert Colour Scanning (HK) Ltd.
Printed in Singapore by Tien Wah Press

Frontispiece: Moschino, 1995.

Contents

'Cunning was well suited to the natural-artificiality that ravished the Victorians: "I love the topiary art," wrote one such, "with its trimness and primness, and its open avowal of its artificial character".
As gentlewomen arranged delicate wax and satin flowers around trompe l'oeil fruit, estate gardeners turned outdoor greenery into cats, swans and Gates of Heaven.'

Hillel Schwartz, *The Culture of the Copy*, 1996

Introduction

Listed as an ingredient of Givenchy's Brilliant Couture lipstick is the word 'nylon'. What could nylon be doing in a lipstick and why are there microfibres in a Helena Rubinstein foundation? One of the great mysteries of the twentieth century has been the arrival and absorption of synthetic materials into our everyday lives; they have moved into our homes, our wardrobes and our handbags, into textiles, clothing and even cosmetics. The silent synthetic revolution has swept through every territory within fashion's vast empire, overturning its traditions and scattering confusion throughout its hierarchical system. Before synthetics, nature's fibres enjoyed an established and secure order as a society of fabrics: linen was 'noble' and silk was 'Queen', but as chemistry's impostors and pretenders became more and more convincing and interesting, designers and consumers began to shift their loyalties.

Never before have so many things been man-made. We are surrounded by plastics and synthetics, from toys to digital technologies and microfibres. Every animal and plant material has been simulated and, some may say, bettered, in the laboratory. So rapid has been the technological progress that we can no longer be sure that the 'natural' material we see and handle is not in fact artificial, and even in the eco-fibre war it is practically impossible to establish whether or not nature's fibres hold the ethical high ground. The story of synthetics is the story of the eternal competition between nature and artifice.

Although we have become, almost unconsciously, totally dependent upon man-made materials, deep within the collective psyche still lingers a suspicion,

Opposite: Perspex corset with butterflies by Alexander McQueen for haute couture Givenchy, 1998. Photographer: Chris Moore.

if not a prejudice, against the words 'plastic' and 'synthetic'. As materials, their reputation peaks and plummets, depending upon their design and cultural context – a nylon Prada bag is a status-plus accessory but a nylon anorak is a fashion joke. It is not what synthetics actually are but what they are believed to be that really counts: synthetics are expressionless materials without any specific character of their own, and their value and meaning is totally constructed by designers, manufacturers and consumers. Identical raw materials can be made into expensive couture collections or cheap mass market clothes. It all depends on the design. In fashion, synthetics have been cast in many roles; nylon was once 'a miracle' and polyester, before it became 'tacky', was a 'wonder fibre'. The infinite mutations of synthetic materials also makes them perplexing and enigmatic. An acrylic might just as easily become a knitted sweater, a painting, a chair or a coffee cup: the oddest relationships are found in the world of the man made.

In turn-of-the-century Paris, Bohemian artists made wooden ties and paper shirts, designed to confuse and subvert the sense of 'normality' in material things, but just as strange and almost as disconcerting was the lingerie made from pulped wood or sheer stockings made from the oily black sludge of petroleum. Synthetics gave us a new world where every material could be fabricated, where a luxurious consumer utopia could be created from everyday things, such as air, water, salt, molasses, limestone, coal or oil. From these unlikely materials, chemists derived thousands of domestic and feminine products: varnishes, paints, polyesters, vivid dyes, fragile nylons and hundreds of synthetic fragrances. In fact, many of the characteristics that make the twentieth century truly modern exist as a consequence of the creation of man-made materials. Celluloid film, which established photography and introduced the movie industry, was made from the same chemical formula as viscose rayon or 'Art Silk'.

Man's conquest of mother nature started with 'semi-synthetics' when cellulose, 'the skeletal stuff of all vegetable life', was rendered soluble through the application of chemicals and turned into fibres or celluloid sheets. Cellulose plus acetic acid then made acetate rayon which was used in everything from clothes to fountain pens, aeroplane windows to lampshades and X-ray film. It had taken centuries of research to invent both the 'mechanical silkworm' and the 'cellulose syrup' that could be turned into fibres and woven into cloth.

Synthetic fibres forced an improbable alliance between three massive global industries, chemicals, textiles and fashion, each of which functioned on conflicting timescales and traditions. In chemicals, the making of explosives led almost accidentally to the making of artificial fibres. Textiles, with its four-thousand-year history, was a slow moving, archaic beast, while fashion was an eager moth drawn to the flame of novelty. The story of synthetics is one of many plot twists and the journey from test tube to clothes rail gave the world a vast new vocabulary of brand names that were chemical tongue twisters. Although there are really only a few generic man-made fibres, including viscose, acetate, polyamide, polyester and acrylic, marketing departments gave the offspring of chemistry and textiles thousands of different aliases. 'Dacron', 'Terylene', 'Trelenka', 'Crimplene', 'Orlon', 'Courtelle', 'Tactel' and 'Tencel' are just a few of the futuristic brand names that Aldous Huxley might well have invented.

At the turn of the century, two great French émigré families dominated the alien worlds of fabrics and explosives. In Britain, Courtauld's had become the world's largest manufacturer of viscose rayon, and in the United States, Du Pont was the world's greatest manufacturer of gunpowder, dynamite and TNT. Du Pont was a chemical giant and the making of explosives ironically led it into the making of nylon stockings. Wilmington, Delaware, the home of the Du Pont company, became the epicentre of the chemical-fibre industry and it was here that the synthetic textile revolution began. Nylon, and almost all the other fully synthetic fibres, originated from work undertaken by Du Pont's research team headed by Dr Wallace H. Carothers. In 1938, the company announced the creation of nylon, the world's first totally synthetic fibre, and when prototype nylon stockings were demonstrated at the New York World's Fair the next year they created a sensation. Here, it seemed, was proof of a future of indestructible luxury for all, of a fashion utopia that could be created in the laboratory. Du Pont's 1939 Children of Science exhibition included many chemical fashions which were described as 'only being equalled in the fictions of Jules Verne or H. G. Wells'.

Fashion, science fiction and world politics came together in the euphoric announcement of man-made

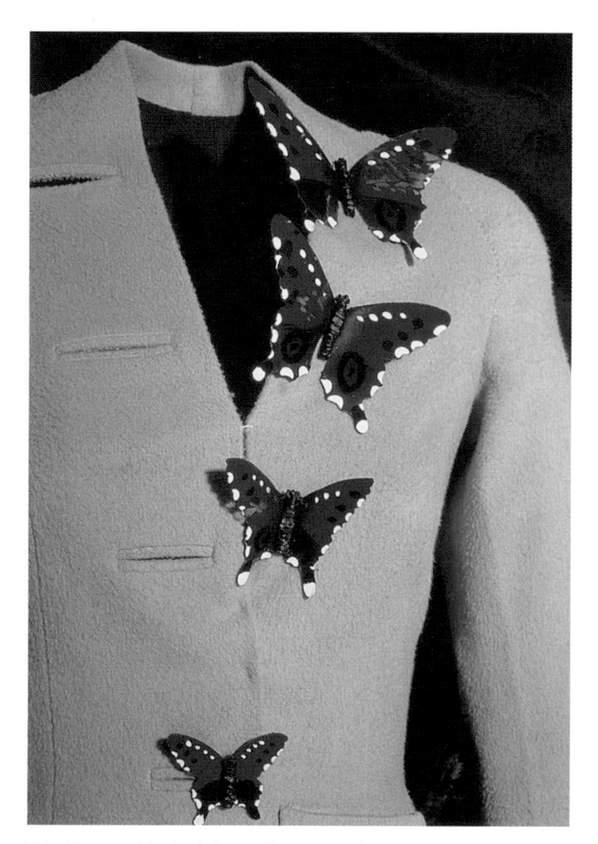

Wool jacket with plastic butterflies by Elsa Schiaparelli, 1937, from the Louvre Museum, Paris.

silk. This fabric, along with synthetic rubber, chemical dyes and other man-mades, seriously challenged the established hierarchy of fabrics and materials. Du Pont's Wonder World of Chemistry fed a popular excitement about the future, for the time when synthetics would simultaneously end female drudgery, invent a consumer paradise and, of greatest national consequence, free America from its dependence on foreign raw materials – particularly Japanese silk. Japan was then the world's largest supplier of pure silk fibre and America was its largest consumer: in reality, Du Pont's nylon stockings were a declaration of economic war on Japan.

Apart from the economic textile revolution brought about by synthetics, design, too, would never be the same again. History has ignored the fact that nylon, polyester, and acrylic were swiftly taken up by couturiers such as Dior, Givenchy and Balmain, but the real fashion potential of the new synthetics was unleashed during the 1960s when Parisian couture found its inspiration in outer space and fashion became an essential element of London's anarchic, throwaway pop culture. Whether or not synthetics have 'class' or 'fashionability' depends upon the prevailing image of science and technology, and after the heady moonstruck excitement of the sixties came the technologically troubled seventies, when man-made fibres slid rapidly into fashion oblivion.

The status rehabilitation of synthetics began gradually when sportswear spilled over onto the fashion pages of the 1980s. 'Lycra' and performance fabrics began the shift but the real restoration of synthetics happened when technology became a nineties 'trend'. Much of today's popular culture is inspired by the amazing breakthroughs in science, on gene splicing, cloning, 'smart' materials and artificial intelligence. Films, books and magazines plunder science facts and shape them into science fiction and, inevitably, the fashions of the future are made with high-tech materials – the 'clever', reactive descendants of nylon and polyester. Alien fabrics, designed for defence purposes or space travel, are constantly appearing in designer fashion collections: a bulletproof miniskirt in 'Kevlar' or trousers with a 'Teflon' finish. Everything points to the fact that fashion has become reinfatuated with technology and the next big question is whether or not the electronic wardrobe and the 'digital' dress will become a whole new garment genre or will the techno-dream fade away and become just another fashion fad?

The history of synthetics is full of paradoxes. Silk is now a cheap 'third-world' fibre while synthetics can cost £300 ($500) per metre. The ingenuity of fibre engineers has made it very difficult to distinguish artificial from natural fibres, and textiles, once an agricultural-based industry, has become largely chemical. Synthetics also made twentieth-century fashion what it is now, a mass-market, fast-moving industry; but where do they stand on the ladder of design status? Are they social 'miracles' or spoilers of the environment? Did they create a completely new cultural and design aesthetic or are they simply cheap surrogates? Did they bring the dream of plenty or a wasteland of worthless things? Both sides of each argument have been voiced at various times and one way of answering such questions might be to imagine another science fiction scenario – one where a strange new virus evolves which attacks only plastics and synthetics. What kind of society would survive the melting away of man-made materials and would we want to live in it?

1

'Thousands of years ago, a humble little white-winged moth laid its eggs on a mulberry leaf and started a monopoly on luxury. Its larvae spun cocoons of a wondrously fine silken filament which patient Chinese learned to strip away.
With painstaking care, the gossamer threads were thrown, twisted and wound on reels, later to be woven into the exquisite fabrics which, for centuries, were worn only by persons of rank or fortune.
In the hope of bringing luxury within reach of all, research chemists – the bold explorers of this new era – set out upon a daring safari in search of a fiber with entirely new characteristics and superior to any yet known that could be synthesized from abundant, low-cost raw materials. Rayon was the "happy" find.'

Du Pont Magazine, September 1941

1700s–1930

The Chemist Conquers the Worm

The story of the centuries-long quest to find a chemically-created substitute for silk is a tale of transition – from the old world, where scientific experimentation was associated with magic and alchemy, to the modern perception that all matter can be rationally understood and scientifically re-created. To turn something base into something precious was the holy grail for most would-be alchemists, and silk (which for thousands of years had been the world's most expensive and mysterious fabric) was as illusive as gold. As is frequently the case in scientific history, chance and accident played their part in the eventual discovery of 'artificial silk' (or 'art silk'). Many and varied were the early chemists whose combined efforts eventually made a viable semi-synthetic fibre, although most had no thought whatsoever of textiles when they were conducting their experiments.

The merging of experimental chemistry with textiles had a slow and meandering start and, ultimately, it was due to the business sense and creative innovations of two great French émigré families that man-made fibres came to revolutionize twentieth-century clothing. Starting from very

Opposite: Antoine Lavoisier, the inventor of modern chemistry, with his wife, Marie Anne Pierrette Paulze. Jacques-Louis David painted this portrait in 1788 and signed the couple's death warrant in 1790.

Above: The dream of ancient alchemy was to transform matter: gold from base materials. The making of artificial silk was just as illusive.

Opposite: After the First World War, Du Pont enjoyed a boom in chemical invention and production.

different manufacturing traditions, the companies of Du Pont and Courtaulds became world-famous names in man-made textiles. Without an understanding of the history of these two families and the historical context of their innovative outlook, it would be difficult to appreciate fully the complicated causes and effects that united chemistry, business, textiles and fashion in the creation of synthetics.

Du Pont: The American Chemical Giant

Du Pont's history is inextricably linked with the origins of modern chemistry and in particular with the French scientist Antoine Lavoisier. Lammot du Pont wrote in 1944, 'In the eighteenth century, chemists, with few exceptions, were little more than apothecaries. And chemistry, instead of being a great science, was scarcely a science at all. It was shot through and through with contradictions and prejudices. Its thinking was thoroughly medieval. It was a hodge-podge of mysticism and magic.'[1] Lavoisier was to change all that: he was the first to turn chemistry into an exact science, sparing neither time nor money to obtain the equipment for his revolutionary quantitative methods of analyzing matter. The du Pont link with Lavoisier laid the foundations for the enormous chemical corporation that was to come.

As superintendent of the French government's gun-powder-making plant at Essone, Lavoisier appointed the young Eleuthère Irénée du Pont, son of his old friend Pierre Samuel du Pont, as his apprentice. Eleuthère was a scientist 'by instinct and by nature' and his passion was to acquire a complete knowledge of chemistry. It was from Lavoisier that the founder of the Du Pont company learnt the techniques of making fine gunpowder.

Lavoisier's great text, *Traite Elementaire de la Chimie*, was published in 1789, the year of the French Revolution. The instability of politics during the years of the Terror made France a dangerous place for all intellectuals. Lavoisier and his wife suffered the harshest fate, succumbing to the cruel guillotine in 1790. (The celebrated artist Jacques Louis David painted the aristocratic couple in 1788, only later to sign their death warrant in his role as a member of the Revolutionary Convention.) 'France', declared Robespierre, 'has no need of scientists.'[2]

Pierre Samuel du Pont, a writer and publisher, unwisely championed the cause of the monarchy. When his home and printing plant were wrecked by revolutionary mobs in 1799, he decided to abandon both France and the family estate at Nemours.[3] The family set their sights on the New World; what better destination for survivors of a bloody revolution than a newly born republic. Among their possessions, Pierre Samuel du Pont brought his library of 4,000 books (said to have cost more to transport than his entire extended family). Landing in New Jersey on 1 January 1800, the du Ponts were to become one of America's leading families and Du Pont one of its leading companies.

The du Ponts were well connected. Thomas Jefferson was a family friend and had visited their estate at Nemours. He encouraged and assisted E. I. du Pont in establishing a business for the manufacture of gunpowder in 1802. America at that time was described as being 'lusty, crude and exacting',[4] and yet a place 'practically without means within itself of supplying gunpowder and explosives. It had no mills or manufactories organized for the purpose of making them, and the little that was made was on a hand-to-mouth principle, just in the same way that people used to make their own leaden bullets in their own homes. On looking around, the du Ponts saw the magnificent opportunity afforded to them of manufacturing explosives of a character and quality entirely unknown to the people here, far ahead of the kind of anything that had been produced'.[5]

The Brandywine Powdermen

Appropriately, America's first great industrial empire located itself in George Washington's first state, Delaware. Perched on a steep, blue, granite hillside overlooking Brandywine Creek near Wilmington, the French-style home built by E. I. du Pont in 1803 still stands. In keeping with his philosophy of sharing risk with his workers, it looms over the former site of his black powder-making mills, and so is erected literally on top of his own powder keg. From here, the du Pont family witnessed and experienced the window-shattering blasts that came from the frequent explosions below. On these terrifying occasions the ladies of the house would rush out into the fields in the hope of escaping the usual secondary detonations.

The Du Pont company started out by employing 140 men (mostly Irish immigrant workers) who lived with their families in a tightly knit community around the powder mills. Extending for over two miles, the powder-making buildings were spaced well apart to reduce the damage when explosions occurred and only two men at a time worked in each one. From the outset, safety regulations were very strict. Matches were absolutely forbidden anywhere near the plant, as was spark-generating metal, to the extent that powdermen could use only wooden buttons to fasten their clothes and wear shoes made with wooden pegs rather than nails; horses, too, wore wooden shoes.

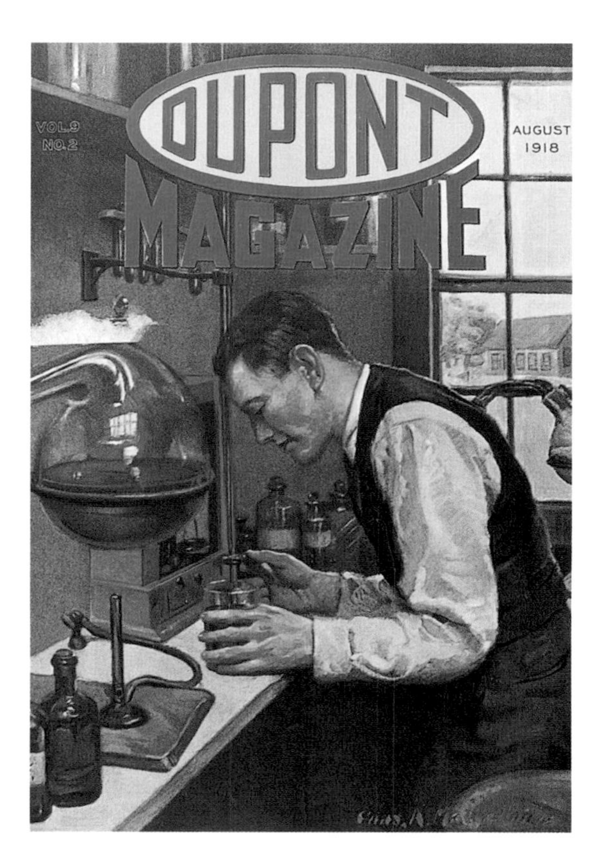

During the nineteenth century there was a limitless market for high explosives, a demand which was boosted by Alfred Nobel's discovery that nitroglycerin could be combined with collodion cotton to form a highly explosive gelatin named dynamite.[6] All this came at a time when the United States was, literally, exploding westwards, blasting out its future through mining and through mountains – laying railways, roadbeds, tunnels and city foundations. It was also a very hostile century in America; battles raged between the settlers and the native Americans, the Civil War divided the country and the frontiersmen made a career out of shooting animals for both food and skins. Du Pont was perfectly placed to become the world's largest black-powder manufacturer, evolving into a corporate giant with its production diversifying into dynamite, shells, bullets, Remington shotguns, and an infinite variety of other domestic products. It left no opportunity untried and also sponsored clay pigeon-shooting.

From its earliest days, the Du Pont company understood the importance of projecting a positive image for its products, and explosives were no exception. In 1911 an Advertising Department was started and in 1916 a Publicity Bureau. Material published during this period points to the diverse and peaceful uses to which explosives might be put: 'Farming With Dynamite' was one suggestion.[7] Trees could be swiftly felled, fields 'ploughed', and a quick way to smoothly split a log was to simply blast it asunder with powder. Ditches, too, could be conveniently excavated with dynamite and it was highly recommended for the planting and cultivation of fruit trees. The company actively portrayed itself as being an explosive manufacturer whose products were intended to enhance rather than extinguish life. Its own company history, published with cruel irony just two years before the First World War, states:

'It is the steadfast purpose of this enlightened age to do away with warfare.... Probably there is no

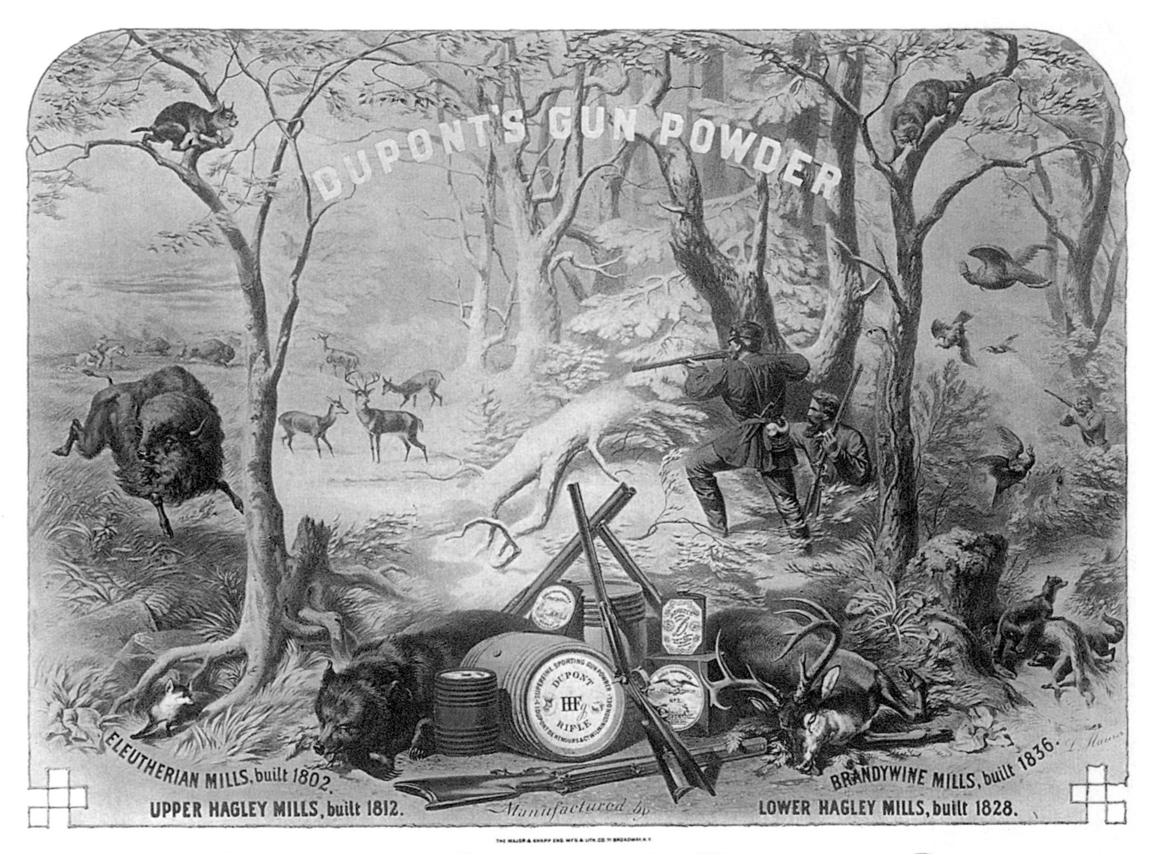

greater force which makes for peace than the invention of high and powerful explosives, for warfare is being rendered so deadly that men realize more and more the necessity of peace.... At the present time not more than 10 per cent of the Du Pont output is used for military purposes. By far the greater bulk of all explosives are used for strictly commercial requirements.... There are no greater civilizing forces than dynamite and blasting powder.'[8]

In 1913, more than 95 per cent of Du Pont explosives went on non-military uses, but within a year 'Du Pont smokeless powder facilities were swamped by frantic Allied pleas for more powder to meet well-equipped Germans.... Employment in the company's own plants rose from 5,500 in 1914 to more than 55,000 in 1918.'[9] By 1916, General Hedlam, chief of the British Munitions Board, was crediting the Du Pont Company with 'saving the British Army'.[10] By the end of the war, Du Pont had supplied over 40 per cent of all the standard explosives used on the Allied side, plus millions of pounds-worth of specialized explosives, caps and fuses. Inevitably, company fortunes flourished as the demand for explosives accelerated.

THE SATURDAY EVENING POST
Inside .. Outside
MADE BY THE MAKERS OF DUCO

Above: The homely, domestic face of Du Pont's chemical products came increasingly to the fore after the First World War. Opposite: Guns and explosives tamed wild America. A late nineteenth-century advertisement for Du Pont's gunpowder.

A Meandering Path to Man-mades

Even before the war, it is clear that Du Pont was not entirely happy with its role as a munitions supplier and preferred to emphasize a more domestic function for its products. Herein lies the seeds for its corporate shift into the first man-made fibres. The seemingly incongruous link between explosives and textiles is, in fact, a logical one: the same raw materials – and so chemical companies – produced both bombs and lingerie fibres.

A century ago, chemistry was inventing a whole new world of products; in 1888 George Eastman introduced the first Kodak camera using film negatives, while the 1890s saw the beginnings of motion pictures, both thanks to celluloid film. Many of these materials had chemical affinities with the first cellulose fibres, but the organic chemists, research scientists, professional inventors and amateur physicists whose combined experiments led to man-made fibres, had no inkling that their dabblings would ultimately revolutionize the giant textile industry. Through a multiplicity of seemingly unrelated discoveries came the series of historical breakthroughs that would endow the world with all the synthetics and plastics produced by the modern petrochemical industries. One prominent triumph for nature's copyists was the invention of the semi-synthetic fibre 'viscose rayon', or 'artificial silk'. The concept of artificiality, a counterfeiting of the authentic, permanently coloured public opinion of all man-made materials and is an issue that troubles the producers of synthetic textiles to this day. In his history of Courtaulds, D. C. Coleman writes: 'Artificial silk was not the long-awaited outcome of a conscious search to find a substitute for silk. It owed much to professional chemists working on the chemistry of cellulose; to the demands of the seemingly quite unrelated industries of electric-lamp manufacture, paper-making, and explosives; and to the energy and ambitions of one particular, and rather eccentric, scientist. Save in the final stages of its development, it owed hardly anything to the existing textile industry.'[11]

Nature's Secret

Before the late nineteenth century, textile fibres consisted only of organic naturals, of wool, flax, cotton, hemp or silk, but from mid-century there was a conscious push to create artificial fibres. The population boom had much to do with an increased demand for fabric, while the technological advances of the industrial revolution made it possible to produce more and more cloth in the expanding textile-weaving factories. The women's fashions of both the eighteenth and nineteenth centuries, such as crinolines and bustles, were designed to consume vast quantities of fabric, which was particularly advantageous to France and England whose economies were largely dependent on the making and exporting of textiles.

Viscose Rayon or 'Artificial Silk' dates from 1885 and was the first man-made fibre to be invented. It was made from cellulose which is the vital component of all plant tissues. More than three centuries ago the English naturalist Dr Robert Hooke anticipated the creation of chemically produced fibres. In his book *Micrographia*, published in 1664, he described how it should be possible to make 'an artificial glutinous composition, much resembling, if not full as good, nay better, than that excrement, or whatever other substance it be out of which the silkworm wire-draws his clew'.[12] A humble worm held the secret of the world's most luxurious fabric, the 'Queen of Fibres'. Of all the natural fibres that have been woven or knitted into cloth, none has the resonance of silk. A much-coveted luxury for many centuries, the source of the fibre and the method by which it could be spun and woven was shrouded in secrecy by the Chinese for 3,000 years after its discovery around 2600BC. The finest and strongest of all the natural textile fibres, it is the solidified protein secreted from two pairs of silk glands (or spinnerets) below the mouth of a silkworm, the larva of the moth *Bombyx mori*. The viscous liquid hardens on contact with the air. The worm then winds and winds the fibre around itself until, after two or three days, the caterpillar is entombed in silk. Remarkably, two kilometres, or more than a mile, of silk thread can be drawn from one cocoon. The caterpillar's sole diet is the white mulberry which is native to China, thus the unique position of the Chinese in silk's early history. The greatest puzzle is how they discovered its potential in the first place. One improbable but often quoted legend has it that a cocoon accidentally fell into the tea

Opposite, top: The beauty and complexity of silk weaving and embroidery in the eighteenth century. Three brocades using silk, silver and gold from Venice and Lyon, c.1730–60.
Opposite, bottom: The caterpillar of *Bombyx mori* or the silkworm.

Above: Eighteenth-century fashions were advertisements of personal wealth and social status. Voluminous dresses consumed vast quantities of luxurious silks. *The Family of Eldred Lancelot-Lee*, 1736, portrait by Joseph Highmore.

Empress Hsi-Ling-Shi was drinking in her garden and, the fibres being moistened and loosened in the hot liquid, they came away in a delicate continuous strand as she drew it from her cup.[13]

The Silk Route was established more than 4,000 years ago. Some 5,000 kilometres (3,000 miles) long, it began in China, ran through Asia and ended at the Mediterranean, from where the trading boats carried the silks to the rest of Europe and the New World. Historically in Europe, silk was the cloth of emperors and queens, of the nobility, the papacy and the clergy. Yards of fabric and elaborate embroidery expressed power and status as clearly as grand architecture and countless Sumptuary Laws were enforced to prohibit commoners wearing silk. The Chinese had jealously protected the mysteries of silk making for thousands of years until, in perhaps the first great act of industrial espionage, the secret of sericulture was finally stolen from China some 2,000 years ago.[14] Within a few centuries, silk technology had spread all over Europe; to Spain by the eighth century, Italy by the twelfth, England and to France, where silk making was established at Lyon in the fifteenth century. Italy became the leading European producer of luxury fabrics until the sixteenth century when French materials of the same quality were produced.

As the silk trade became more and more lucrative, Dr Hooke's original idea of making a synthetic glutinous thread re-emerged. Chemists attempted to recreate lustrous silk fabrics but at a price that would be within the income of the many rather than the few. It is hard to imagine the countless years spent by dozens of scientists, observing the lifespan of the silkworm, trying to emulate its digestive processes and striving to imitate the cellulose of mulberry leaves. What were the chemical secrets of reconstituting the regurgitated organic material and how did it become so delicate and yet so strong?

From Syrupy Solution to Solid Fibre: the Chemistry of Artificial Silk

Cellulose is nature's polymer and consists of hundreds of molecules strung together like a beaded necklace. It is derived from chemically pulped-down woody plants, such as the fast-growing trees spruce, hemlock and pine, or from cotton linters that are then regenerated. Both viscose rayon and paper are made from cellulose and progress towards the creation of the textile fibres in the nineteenth century was assisted by improvements in paper manufacture. However, the conversion of wood cellulose into textile fibres is difficult because the cellulose polymer chains cling firmly together and the only way to disentangle them is to dissolve them in caustic soda and other chemical solutions. A German weaver, F. Gottfried Keller, discovered the chemical process that would dissolve wood pulp in 1840. As in so many instances of scientific discovery, it was an almost chance observation – he watched his children grinding cherry stones and noticed that the powdered wood floated on the surface of water; when he squeezed this pulp in his hand it acquired the same consistency as rag pulp. He had taken the first step towards making the raw fibrous material that would eventually become rayon.

During the early part of the nineteenth century, several European chemists set about treating cellulose in its many forms – wood, cotton, paper, linen – with various acids. Among their findings was the important but alarming discovery in the 1840s by Friedrich Schoenbein, a Swiss scientist, that the result of treating cotton with nitric acid was the highly explosive, nitro-cellulose substance known as gun-cotton. From this discovery stemmed both the modern explosives industry and the process of making artificial silk. When Schoenbein's nitrocellulose was mixed with camphor it produced celluloid. This was fully developed into a nitrocellulose form of fibre by another Swiss chemist, Georges Audemars. He discovered that, 'If he dipped a needle into a solution of nitrocellulose and then drew it away, a filament was formed which dried and hardened in the air and could be wound up into a reel.'[15] Audemars' was the first actual and British patent for artificial silk, taken out in 1855 and describing 'improvements in obtaining and treating vegetable fibres', but no textile interests were aroused. In the mid-nineteenth century, patents for fibre-making came thick and fast. Some were odd concoctions, including the mixture of fat, glue, gelatin, oil, wheat and cellulose registered by E. J. Hughes in Manchester.

Another Englishman, the physicist and chemist Sir Joseph Wilson Swan and the inventor of the first artificial light filaments in 1883 (later to be used in Edison's electric bulbs), needed to create a flexible fibre of uniform thickness to carry the electric current. His method was 'to squirt nitro-cellulose, emulsified in acetic acid, through a small die into a coagulating bath of alcohol, so that it formed into threads of indefinite length, being afterwards denitrated'.[16] Struck by the potential of fine-diameter filaments for textiles, Sir Joseph had some specially created threads crocheted into mats which were exhibited at the Exhibition of Inventions in 1885 under the description 'artificial silk'. His prime interest, however, was in producing a filament for his newly invented electric lamp and nothing more came of this tentative alternative to traditional textiles.

Crucially, it was when the nitrocellulose process was taken up by the Frenchman Count Louis Marie Hilaire Bernigaud de Chardonnet in 1892 that the task of perfecting artificial silk was achieved. Chardonnet was given financial assistance not by the French textile industry but by a paper-maker. His was the first textile-intended man-made fibre and was specifically described in his 1885 patent as *'une matière textile artificielle ressemblant à la soie'*. Coleman acknowledges Chardonnet as 'the father of the artificial silk industry' because his discovery 'really was the result of a conscious personal search, for *"soie artificielle"*; and he followed up his initial invention by becoming the first man to set up an artificial silk company, to start a factory for its manufacture and to reach successful commercial production in this new fibre.'[17]

Chardonnet had been trying to discover an artificial fibre since 1878. As a student he had studied under Louis Pasteur in Paris and knew of his efforts to cure the infamous caterpillar disease *Pebrine* which at one time had threatened to completely destroy the French silk industry. Aside from jeopardizing silk producers' livelihoods, factors such as disease led to dramatic fluctuations in the price of raw silk. Chardonnet set out to copy chemically the silkworm's spun silk. At first he came up with nitrate silk, a highly flammable and explosive form of rayon, but in 1884 he created his first artificial fibres from a nitrocellulose solution which was squirted through a spinneret (an apparatus with tiny holes, similar to a shower head).

Louis Marie Hilaire Bernigaud, Comte de Chardonnet, the pioneering inventor of '*soie artificielle*'.

The fibres then hardened in warm air. This was the first commercially useful, multi-filament yarn and the French paper-making firm, J. P. Weibel, supplied the capital for Chardonnet to open a factory at Besançon. Materials made from Chardonnet Silk were exhibited at the International Exhibition in Paris in 1889 and attracted much attention from enterprising textile firms. Several factories were opened under licence in various parts of Europe and by 1898 he had begun to make a profit.

The essential flaw in the new material soon manifested itself. Because they were not denitrated, the fibres were highly flammable and volatile, so much so that some early experiments, and indeed factories, came to an explosive end. Explosions occurred at Besançon in the early 1890s, at Tubize in 1904 and later at a Hungarian factory. Sometimes Chardonnet's rayon was disparagingly referred to as 'mother-in-law silk', presumably because a bolt of this fabric would make an ideal present for a troublesome mother-in-law who might obligingly sit next to an open fire or gas flame. These nitrocellulose fibres were, in fact, bombs in fabric form and some English textile journals issued grave warnings about the wearing of materials made from such dangerous filaments and the French Government forbade their manufacture.[18] This was a potential disaster for Chardonnet and he was only able to save the future of his company by adopting Sir Joseph Swan's denitration process. Chardonnet Silk, based as it was on nitrocellulose, remained a hazardous product to manufacture and its production was only ever sporadic. (The last remaining Chardonnet Silk factory, located in Brazil, succumbed to a final, inevitable, blaze in 1949.)

Viscose Rayon

The final breakthrough in the 'viscose' process was made by professional British chemists. In 1892, a master patent for the making of a viscose compound was taken out by Charles F. Cross, Edward J. Bevan and Clayton Beadle whose work on the chemistry of cellulose was stimulated by their research into the paper-making industry. Their patented method was to treat wood pulp with caustic soda and other chemicals to produce a golden yellow substance which they named 'viscose'. Cross and his partners discovered that it was possible to dissolve cellulose without first making it into nitrocellulose. They found that if any cellulose material (whether obtained from wood, flax, cotton or other raw materials) was broken down with caustic soda, it could be treated with bi-sulphide to create a highly viscous solution. This solubilized viscose was then forced through jets into an acid coagulating bath to produce the fibres; if forced through a narrow slit it emerged as cellophane. The acid neutralized the material and made it insoluble.

Another patent for an acetate process was taken out by C. F. Cross in 1894 and, not surprisingly, with his background in the paper industry, it was a method for converting cotton linters or wood chips into sheets of pure cellulose. The cellulose in this case was rendered soluble in acetone and, as a material, acetate rayon became a close rival to viscose. Cross's work was furthered by the Swiss brothers Henri and Camille Dreyfus who first used their version of acetate as varnish or dope with which to paint the wings and fuselage of biplanes.[19] In 1916, the British government invited them to England and sponsored a factory to manufacture their cellulose acetate for this very purpose. At the end of the war, however, the Dreyfus brothers had to find another end-use for their large dope-producing factory and turned their attention towards the development of a cellulose acetate yarn. By 1921 they had overcome most of the technical problems and they were making an acetate fibre which they named Celanese.[20]

The Dreyfus brothers' process was perfected by C. H. Stearn (scientist, inventor, electric-lamp manufacturer and former collaborator with Swan) and C. F. Topham (glass-blower, engineer and employee of Stearn) who patented a method of turning viscose into fine filaments. Cross and Stearn then collaborated in setting up the Viscose Spinning Syndicate in 1898 in a laboratory at Kew, and by 1902 the Viscose Development Company Ltd had been formed. Their joint purpose was to make lamp filaments and 'artificial silk fibres'. By 1904, all the national patent rights had been sold: 'to the little works at Kew came Germans, Americans, Frenchmen, Russians; and Englishmen, too, some of them from textile firms at last aroused to the possibilities of this new invention.'[21]

Another method for creating semi-synthetic fibres was the 'cuprammonium' process. Developed by the chemist Louis Henri Despaisses in France in 1890, the cuprammonium process at first failed to compete with the more economical viscose production, but it was successfully revived by the German company J. P. Bemberg in 1919 and was used to produce a very fine denier fibre.[22]

The name Cuprammonium derives from copper sulphate and ammonia which, in conjunction with caustic soda, are used to solubilize wood pulp. The European textile industry began to take an interest in the new 'imitation silk' processes, Cuprammonium – and in spite of its dangers – Chardonnet's method, and many factories opened in Britain and on the Continent. Initially, artificial silk was very weak when wet and came apart at the seams and it was not until around 1910–14 that improvements led to more reliable fabrics being produced, although they still had to be washed with great care.

Courtaulds' Rayon

By the turn of the century, conditions were ripe to change the textiles industry for ever: the scientific processes for making 'imitation silk' were resolved and the most outstanding textile makers in Britain, Courtauld's, were in search of a new direction and a new product. The Courtauld's name was soon to become synonymous with the production of viscose rayon and the company's venture into semi-synthetic fibres blazed a new textile trail into the twentieth century. It was to become the archetypal success story of a once great company facing a decline but restored and made even more profitable by embracing a completely new technology.

Like Du Pont, Courtauld's began as a family firm. The Courtaulds arrived in England in the late seventeenth century as Huguenot refugees and their involvement with British textiles began almost immediately. A century later, George Courtauld I, the father of the company's founder, was apprenticed to a silk throwster in the famous Huguenot weaving settlement in London's Spitalfields. He was managing a silk-throwing mill in Essex by the beginning of the nineteenth century and in 1809 set up a mill of his own. His son, Samuel Courtauld III, established the famous silk weaving business in 1816, in collaboration with a series of brothers, sisters and cousins.

The Courtauld's Textile Company owed its prosperity to the nineteenth-century fixation with death and its elaborate rituals, in particular the lengthy period of mourning when Victorian etiquette demanded that women should be shrouded for months, if not years, in black. In 1830, Samuel Courtauld began the manufacture of mourning crape (a black figured silk, gloomy enough for deep mourning) and the company soon became its largest and most successful supplier. After 1850, this fabric dominated Courtauld's manufacturing output. The company history describes the years between 1850 and 1865 as 'halcyon, both for mourning crape and for profits'; 1835 to 1885 the firm's annual profits rocketed from £3,000 to £110,000.[23] New mills were built or were acquired and the company employed 3,000 people. Periodic booms followed royal deaths and profits improved when the French began to acquire a taste for 'crape anglaise'. Peter Robinson opened the vast Court and General Mourning Warehouse and countless retail outlets had special mourning departments. The social requirement of wearing the black of the bereaved kept prices high while the relative secrecy around the crimping techniques of crape-making – Courtauld's employees were sworn to secrecy – ensured that Samuel Courtauld & Co. (the company name from 1828) had few competitors and dominated the business.

The popularity of formal crape-laden mourning began to wane towards the end of the century and the

Above: Mourning etiquette was a Victorian obsession. The amount of black, matt, silk crape worn (here, covering much of the skirt) was in direct proportion to the recency of the bereavement.
Below: Courtaulds became an enormous textile empire based on the success of its fashionable mourning crape.

Something New
IN
Fashionable Mourning:

COURTAULD'S
Waterproof
Crape

CAN BE HAD
(1) In the usual firm finish,
OR
(2) in a soft finish.

The latter is particularly suitable for trimming soft dress materials.

FROM LEADING WAREHOUSEMEN.

crisis came after 1885 when prices fell sharply, suffering a 40 per cent drop in less than ten years. Profits tumbled and in 1894 the company made a loss. The final death blow to the mourning crape business came with the demise of Queen Victoria, the monarch who had made a virtual fetish of mourning, in 1901. This was a critical time for Courtauld's but they were quick to respond to the social changes and developed completely new fabrics for totally different markets, beginning to manufacture light dress fabrics which they named crepe Italien, crepe de Pekin, crepe Espagnol, crepe Indien and crepe de Chine – the last of which became a classic fashion material.

Around this time, Courtauld's was reorganized, with the declared intention of manufacturing 'crape, silk, wool, cotton and other fibrous substances of all kinds'. The key figure and engine of change within the company was a Yorkshireman called Henry Greenwood Tetley. Although he was not a member of the family, Tetley and his fellow board-member, Thomas Paul Latham, virtually controlled the company during the next quarter of a century, beginning in 1904 when Tetley told the board that the firm needed 'a new source of profit to replace crape profits – which are leaving us'. His remedy was to move the company in a wholly new direction and one which, in a global sense, marked the real launch of the man-made fibre industry.

In July 1904, at Tetley's instigation and insistence, Samuel Courtauld & Co. bought a set of patents and licences (at a cost of £25,000), giving them the exclusive British rights to the 'viscose' process of making 'artificial silk', later known throughout the world as rayon. These were the patents of Cross and Stearn's Viscose Spinning Syndicate. It was a personal triumph for Tetley when Courtauld's acquired the patents as no director of the company at the time knew anything about chemistry. A factory was quickly built near Coventry where the new viscose yarn began to be manufactured in 1905. It was an enterprise not without risk: would the conservative British manufacturers and public accept this new, alien material, and would it be a viable fabric or would the patent-holders in other countries make a more successful viscose yarn? The years 1905 to 1913 saw 'the transformation of a textile firm into a chemical firm with a textile branch.... For already by 1909 the profits which came from the chemical procedure of turning wood-pulp into artificial silk yarn were

Above: In the 1920s, Courtaulds' factories were producing vast quantities of viscose rayon. The Coventry factory, 1928.
Opposite: A 1946 advertisement for British Celanese which produced cellulose textiles, plastics and chemicals.

greater than those which came from making fabrics ...more and more the company's main product was yarn to be processed by textile firms not a fabric to be worn by the consumer.'[24]

Experiments continued for several years to find the best use for the viscose yarns. At first they were made into braids and trimmings but by 1909 they were being used to make ties and, under the trademark 'Luvisca' (fabrics with a cotton warp and a viscose weft), were made into shirts, while silk and viscose cloths were made into ladies' blouses. The use of viscose for making linings was an important advance but the real breakthrough came in 1912 when Courtauld's salesmen persuaded their largest single customer, Wardle & Davenport, that viscose yarn would be appropriate for knitting stockings. Courtauld's appointed special salesmen whose job it was to vigorously push their new fibre; a network of agents at home and overseas also helped to sell their product to manufacturers (it was Courtauld's policy to abdicate responsibility for the woven goods actually made by the firms that bought their artificial fibre). In the end, Samuel Courtauld & Co. had the edge over all its competitors because, of all the purchasers of the viscose patent rights, they alone were textile manufacturers. This meant that:

'...those directing the efforts of the chemists or engineers knew just what technical qualities were needed to make a yarn useful and saleable; the textile machinery and the dye-houses at the Essex mills were available for experimenting with Coventry's products; and the whole range of commercial contacts made by the company in its textile business could be used for the marketing of a new product. One especial goal they knew to be of particular importance: to make a yarn capable of being woven and dyed. Hitherto the other sorts of artificial silk had been really effective only in non-woven uses, e.g. for braids, tassels, fringes or the like, as substitutes for silk. If viscose yarn could become a textile fibre in its own right then its future was assured.'[25]

It did not take long for the Coventry factory to demonstrate the superiority of its viscose-making process over the existing Chardonnet and cuprammonium methods and to make Courtauld's the pioneer of a commercially successful rayon. In 1912, Coventry's profits alone were over £300,000 for the year and just one year later the company's £5 shares were changing hands at £35.[26] By 1913 a new company, Courtaulds Ltd, valued at £2 million, was formed. Tetley's instincts about viscose rayon proved to be right and Courtaulds developed the process into the world's most successful rayon manufacturing business. It became the largest of the viscose patent-owning firms, buying out the holders of the American rights and setting up a wholly-owned subsidiary in the United States, the American Viscose Company (AVC), which began production at a factory in Pennsylvania in 1911. By 1914, Courtaulds had the patent-based monopoly on viscose yarn production in both Great Britain and the United States. When Tetley died in 1921 the company passed back into the hands of a family member, Samuel Courtauld IV (great-nephew of Samuel Courtauld III), who ran the company for the next 25 years.[27] The general public became most familiar with rayon during the 1920s in its cellulose acetate form which in Britain was sold under the brand name of 'Celanese' and was widely used in the manufacture of

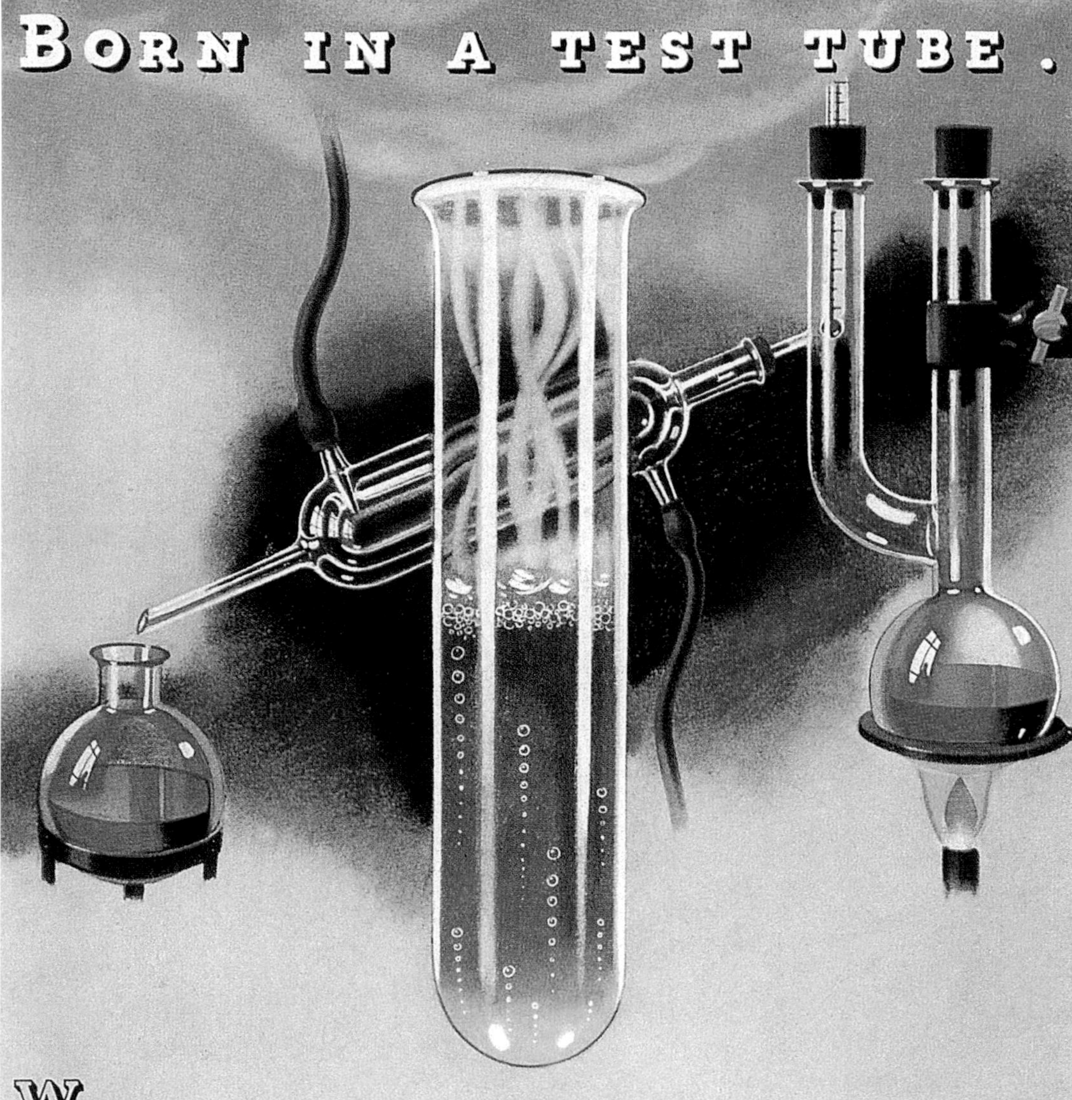

WHEN the Industrial History of this Era comes to be written, one of the most intriguing chapters will surely be headed "The Age of Synthetics". For many years now the test tube has given birth to new and amazing Products, but it took war to lift Synthetics out of the class of mere "imitation" to a level where they are accepted as man-made Products wholly efficient in their own field of application. The crowning triumph of achievement lies in the fact that many modern synthetics are an improvement upon the Product they replace.

British Celanese Limited take pride in the constant development of new synthetic products. The ingenuity of Celanese scientists is ever being called upon to create commodities essential to the betterment of modern industry and social life.

'Celanese' TRADE MARK { TEXTILES PLASTICS CHEMICALS

BRITISH CELANESE LIMITED, CELANESE HOUSE, HANOVER SQUARE, LONDON, W.1

Above: Challenge Cleanable Collars were made from a paper-thin celluloid and muslin sandwich which was then heat-pressed to simulate the texture of linen, 'polished' to give the impression of starching and finished with faux hem stitching.
Opposite, top: One of several prototype hats made in the 1930s for Du Pont's Style Service. This model was easily crocheted by the home milliner from opaque ribbons of cellophane cellulose film.
Opposite, bottom: An advertisement for Courtaulds' lingerie, 1932.

lingerie. The manufacture of 'artificial silk' began an international expansion and a rayon boom was soon under way. Courtaulds built a Canadian yarn factory in 1925, a French plant, near Calais, in 1927 and another at Cologne in 1928.

The United States: Du Pont Enters the Fray

In the United States, the most serious competition to Courtaulds came from the powerful Du Pont rayon-making enterprise. The Du Pont company was already veering towards the concept of chemically-created materials when, in 1910, it bought out the Fabrikoid Company of Newburg, New York and began the manufacture of pryoxylin-coated fabrics as leather substitutes. This was at a time when: 'Cow production could hardly keep up with car production ... so leathermen were finding it difficult to meet with demand for automobile upholstery material. Now, with a huge market waiting in Detroit, Du Pont chemists set out to increase the product's resistance to weather, dirt and wear. They succeeded – then went further. The new "Fabrikoid" resisted grease, oil, perspiration and mildew – enemy agents as far as leather was concerned. The new material was offered in many colors, weights and surface finishes at a fraction of leather's cost.'[28]

'Fabrikoid' was often used for luggage and as a popular imitation leather for women's handbags. Other Du Pont-produced stand-in materials were cleanable collars and an 'artificial ivory' or 'Pr-ra-lin' (made using the nitrocellulose process). These new materials were seen as the means by which life's luxuries could be furnished to everyone, and a patriotic glow of democratic idealism stands out in the advertising and sales material of the day. 'Once upon a time only kings knew the luxury of ivory', runs a feature in the *Du Pont Magazine* in 1918, 'Today, skilled craftsmen versed in the magic lore of chemistry recreate for you and I the exquisite graining and delicate mellowness of those gleaming tusks of old.'[29] It seemed a natural progression for Du Pont to move into the manufacture of artificial silk and in 1920 it bought the American rights to manufacture a form of viscose from the giant French producer Le Comptoir des Textiles Artificels. In 1921, the Du Pont 'Fibersilk' company manufactured its first viscose fibres.

The search for a generic name that would cover the various types of artificial silk became more urgent in the face of complaints like that from one dressmaker who protested, 'Calling viscose Artificial Silk is tantamount to calling a steel beam "artificial timber", or a brick "artificial stone."'[30] In response to these frustrations, the National Retail Dry Goods Association (a department and drapery store trade group) launched an America-wide competition to find a new name for viscose, the only condition being that the new name should not include the word 'silk'. Among the public's suggestions were 'filatex', 'glis', 'glistra' and the recurring 'klis' – 'silk' spelt backwards. Thomas Edison came up with 'plant silk' as distinguished from 'worm silk', but the search finally ended when Kenneth Lord of Galey & Lord's (textile manufacturers) came up with 'rayon', said to be derived from the French for 'shine' and making reference to viscose's main selling feature, its silk-like lustre.[31] The word 'rayon' was formally recognized by the United States Federal Trade Commission in 1925, adopted by the Silk Association of Great Britain and Ireland in 1926 and sanctioned by Courtaulds in 1929.

With a seemingly unlimited domestic market, Courtaulds' AVC produced more than 60 per cent of American rayon. Rayon gave Courtaulds nearly half its gross income in 1927–8 and in the 20 years between the wars, world output of rayon rose from 13 million kg (29 million lb) to more than 1,000 million kg (2,200 million lb), an increase of 7,600 per cent.[32] All this in spite of the fact that, as competition grew, the price of fibre fell sharply. This inevitably brought rayon within the reach of the ordinary woman so that 'cheap woven or knitted fabrics and hosiery, in rayon or rayon mixtures, tapped new social levels of demand: cheaper stockings or underwear, cheaper furnishing fabrics or dress materials'.[33] Ultimately, this downward expansion of rayon's market would give it a damaging down-market image.

A Powerful Weapon: Advertising

The Celanese Corporation of America began producing acetate in the United States in 1924 under the trademark 'Celanese'. Many other manufacturers produced the same fibre under their own trade names. Courtaulds entered the acetate market in 1927 and marketed it as a 'beauty fibre'. These were the chemist's copies of the sleek, luxurious satins and taffetas that were the epitome of twenties and thirties fashion, and the profusion of contemporary advertisements show that the rayon-fibre companies were working hard to promote the new fabrics in a glamorous context. Man-mades became the mainstay of the lingerie and hosiery industries between the wars and public awareness soared thanks to advertising. A survey carried out for Courtaulds in 1930 showed that '35 per cent of a sample of women questioned in London, Manchester and Bournemouth had never heard of Courtaulds but 98 per cent had heard of Celanese; 46 per cent knew that they owned garments made of Celanese, only 6 per cent were conscious of having bought material of Courtaulds' manufacture.'[34]

The arrival of man-made fibres clearly gave a boost to the advertising industry and the purchasing power of newly enfranchised women began to be appreciated and courted by many of the big corporations. Magazines featured a host of feminine advertisements paying tribute to the ladies' new role as purchasing agents for the American family. The 1920s were a

It started one Saturday night
THERE'S no use talking—Saturday was an exciting day for Amos Hunter. (You know him—the nice young fellow with the pink cheeks.) That night he had a date with a girl. THE date with THE girl.
So he spent part of the afternoon shining up the old bus with "Duco" Polish. By suppertime it was new looking and handsome enough for a king and queen.
In the meantime Susie Blossom was busy with her needle, putting the last frills and furbelows on her new, peach-colored Du Pont Rayon dress.
Susie was pretty sure that Amos was THE boy.
And at eight o'clock sharp, Amos and Susie were on their way in the bright, shiny car . . .
. . . bound for the movies to see a hand-holding romance that was made on Du Pont film.
Neither Amos nor Susie realized how chemical research had touched their lives that day. The shiny car, the rayon dress, and the movie film—all resulted from the work of chemists. As a matter of fact, no day passes that modern chemistry doesn't help make life happier and more complete for them—and for you.
BETTER THINGS for BETTER LIVING
. . . THROUGH CHEMISTRY
DU PONT
Producers of Chemical Products since 1802
Listen to "The Cavalcade of America" Wednesday evenings, 8 P. M., E. S. T., beginning October 9th, over C. B. S. network

Above: 'No day passes that modern chemistry doesn't help make life happier and more complete for them – and for you.' *Du Pont Magazine*, 1935.
Opposite: Elsa Schiaparelli's 'glass cape' of 1934. More than any other designer of the time, Schiaparelli embraced the creative potential of semi-synthetics.

commercial turning point, with the 'dramatic mass emergence of many interesting phenomena, including the automobile, the radio, the motion picture, and woman. Most, including woman, had been around for some time: only now did the scene suddenly swarm with them.... Women learned to drive cars, to the delight of cartoonists, and feminine smoking no longer was regarded as wicked. Cigarette sales jumped from 47 billion in 1920 to 125 billion in 1930. Particularly striking was the change in woman's appearance. Skirts rose from the ankle to the knee. Girdles replaced corsets and underwear became firmly lingerie. Up to this time industry had for the most part served woman chiefly by indirection. But now she wanted a larger wardrobe [and] numerous changes of costume to match her speedy pace'[35] – which could readily be supplied in rayon, if the fibre-makers could convince them that rayon was just as desirable as silk.

In 1927, Du Pont commissioned a report on the current standing of rayon and a campaign that would 're-launch rayon as a high status fashion material'. This document gave four key reasons for negative consumer associations with 'artificial silk': the poor quality of the early product remained in people's minds; it was sold as an artificial product, as a substitute, which implied inferiority; manufacturers did not understand the best way to produce it and so quality suffered again; and, due partly to the imperfection of the product in its early days and partly to misuse by manufacturers, artificial silk materials neither washed nor wore satisfactorily. 'The present quality of this material is greatly improved.... Many reputable manufacturers whose own reputations are at stake are now properly using it, so that finished products received by the trade and the public are entirely satisfactory. The introduction of the new generic name "Rayon" has tended to stifle former prejudice which arose from the application of terms such as "artificial" and "substitute"...the situation is steadily improving...but a normal course of correction would require many years and, during the interim, Rayon's progress could not be as rapid as the industry would wish.'[36]

'Men and women should think of rayon exactly as they think of silk as silk, or wool as wool'. It should not be an inferior silk or a glorified cotton but a distinctly separate fabric with qualities all its own. 'Educational propaganda' should be undertaken which would correct the misunderstanding of rayon and the unfavourable impression it made, not just among consumers but the whole fashion trade, including weavers, converters, finishers, sales agents, cutters, underwear manufacturers and retailers. The means by which rayon's reputation could be enhanced was to link it indisputably with high fashion and specifically with the most famous Parisian designers of the day.

The fibre and textile industries were acutely conscious of the harm that was done to new fabrics because of the disparaging prefix 'artificial' in their description and by the association of rayon with the down-market end of the fashion industry. Unlike wool, linen and silk, there was a lack of a design tradition for man-made fabrics and early rayons were often frowned upon by couture designers.

The French couture industry, which still led the fashion world, was in turn led by the French textiles industry – which itself was firmly rooted in the production of luxury fibres. Couture was an unashamedly luxurious market, reserved for the traditional, high status fabrics – but there are always exceptions and even in 1930s Paris there were a few fashion adventures with rayons and cellophanes. Nina Ricci, for example, made a black shimmering evening coat from cellophane in 1935, but it was Elsa Schiaparelli who embraced the experimental potential of the new synthetics with most enthusiasm. With a surrealist's eye for novelty, she accepted no class barriers in fabrics and worked in collaboration with the French textile company Colcombet to develop and to design with the most bizarre and newest man-made materials.[37] In the mid-thirties, Schiaparelli was already using clear perspex, lucite and plexiglass in her accessories, making dresses from tree bark and often incorporating cellophane into her designs. She claimed that ideas always sprang from fabrics and perhaps one of the most radical of these was the 'glass cape' she made from Colcombet's 'Rhodophane', a brittle, transparent spun synthetic. (Colcombet also produced a fluorescent fabric called 'Rhodia Satin'.) It was a risky business making couture garments from these untried chemical fabrics, as *Vogue*'s editor Diana Vreeland found when she sent her Schiaparelli gown to the dry cleaners: it reverted to an oily sludge when it came into contact with the dry-cleaning fluids and ended up in a bucket.

Creating a stylish, designer image for semi-synthetics was a high priority for all man-made fibre companies, and particularly for Courtaulds and Du Pont. Advertising was very much a part of the whole Du Pont business culture, needed both to combat competition and to establish new markets for unknown materials. In 1910, the Du Pont Company's advertising budget was already $250,000 a year and by the outbreak of the First World War, the advertising department was 'as complete and efficient an organization as exists anywhere in the world... with a personnel of approximately 200 men and women ...including a director, assistant director, six advertising managers, an export advertising manager, editorial staff of the *Du Pont Magazine*, a school staff for the instruction of salesmen, an art department, photography gallery, shipping department, and news service bureau.'[38] It was the declared intention of the Du Pont advertising department to 'create the realization by the public that the name Du Pont stood for progress', and they and the other major new-fibre

producers like Britain's Courtaulds and ICI (Imperial Chemical Industries) set about converting consumers to the strange new laboratory-made substances that had even stranger sounding brand names.

Advertising was the way in which the fibre creators could reach both the fashion consumer and their own immediate clients, the textile producers. And the fortunes invested in patenting, branding and marketing were felt to be justifiable because there was widespread confusion about the nature of artificial fibres. Then as now, for consumers and designers alike, synthetics were regarded as a technical mystery – what exactly are they made from? Few could, or can, make the distinction between a cellulose fibre (a semi-synthetic) and a polyester (a true synthetic) or are aware that one is made from a natural source while the other is a product of the petrochemical industry. What unites and defines them all as being 'man-made' is that they are artificially and chemically processed.

Viscose and acetate are *semi*-synthetic – part chemical, part natural – because they are made from pulped organic materials such as wood, cotton or other vegetable matter. Rayon was, therefore, only the first step on the road to real synthetic fibres. It was the half-way stage, a chemically reconstructed fibre that was made from a natural organic cellulose source. Known as bast fibres, almost any fibrous vegetable material can and has been made into fabrics, including hemp, ramie, jute, pineapples, leeks, nettles and seaweed. The raw material, cellulose, is a natural protein found in plant stems but it can also be derived from other sources like casein (from milk), zein (from maize) and arachin (from groundnuts); these too have been converted into fibres but have never really become significant as textiles.

The arrival of man-made fibres was to permanently alter the way in which clothes were to be designed, made, bought and worn. The semi-synthetics were used mostly for lingerie and hosiery during the 1920s and 1930s and gave at least the appearance of 'luxury' fabrics for all. They marked the beginning of a gradual democratization of fashion, a fine aspiration that was voiced almost a century earlier by Peter Alfred Taylor (then a partner in the Courtauld's silk firm) in a speech to celebrate the firm's twenty-first anniversary in 1846. He could foresee 'no reason at all why the luxuries as well as the comforts of life should not be brought home to the cottage of the artisan. Which shows that what one age terms luxuries may in the next age become by common usage the undisputed right of all.'[39] 'Three hundred years ago', he commented, 'the first pair of silk stockings was worn by Queen Elizabeth', and, by the nineteenth century, the hose of royalty was being worn by commoners. Exactly one hundred years later, he would have been astonished to see the frenzy that nylon stockings generated and delighted by the access to fashionable dress made possible by the other post-war synthetic fibres, polyester and acrylic.

The fashion designer had little visible role in the story of semi-synthetics. It was a three-part drama between the research chemist, the fibre/textile manufacturer and the consumer, scripted by advertising agencies in a propaganda war that was designed to propel popular opinion towards the promised chemical-clothes utopia. It is the incongruous juxtaposition of the polymer chemist and the lingerie department that makes the period up to the Second World War so bizarre and sometimes so comical in terms of advertising imagery, but for the first time, these two quite different cultures were drawn into a necessary alliance. As the explosives and 'art silk' textiles industries melted into one another, the wonder world of synthetic materials was soon to make its full social and design impact, and the battle of the brands was about to accelerate with the arrival of the first fully synthetic fibre – nylon. Nylon was to be the detonator of a massive advertising explosion and one that would be targeted primarily at women during the 1940s and 1950s.

Man-made fibres brought a cataclysmic change to three enormous global industries: chemicals, textiles and fashion, with their conflicting traditions and economic pressures. The interface between these three great commercial forces gained momentum as the full synthetic revolution began to unfold. Although dozens of European chemists, physicists, and inventors had, over many centuries, laid the research foundations for the man-made fibre industry, it was the American Du Pont Company, the giant of world chemical research, who gave the world the first completely synthetic fibre and whose chemists developed almost all of this century's synthetic fibres.

Opposite: The synthetic continent created by Ortho Plastic Novelties and published in *Fortune* magazine in 1940. The countries emerge from the natural world of plants into the territory of the molecule and the points on the chemical compass are carbon, hydrogen, oxygen and nitrogen.

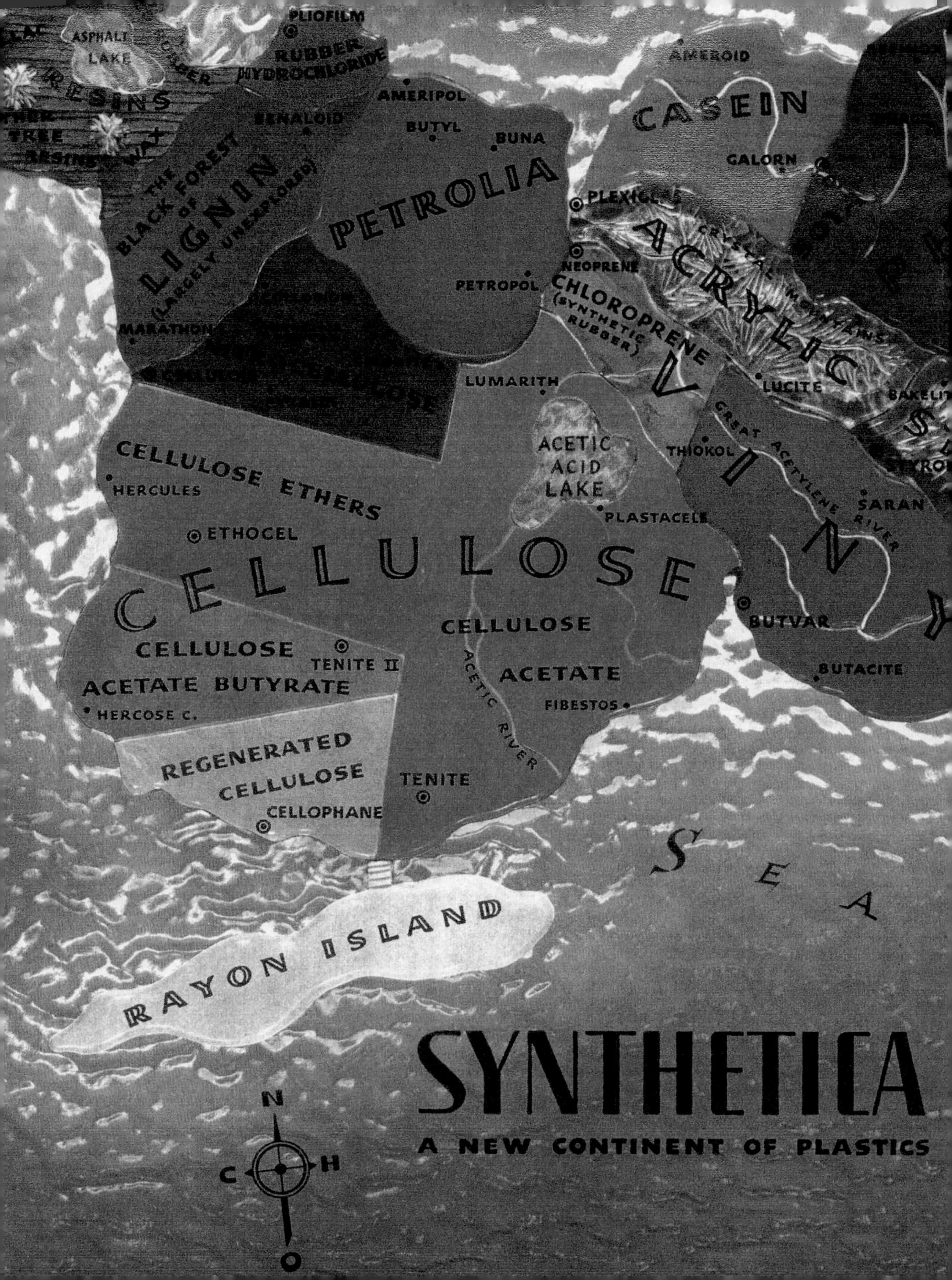

PLIOFILM
ASPHALT LAKE
RUBBER HYDROCHLORIDE
RESINS
TREE
AMEROID
CASEIN
AMERIPOL
BENALOID
BUTYL
BUNA
GALORN
THE BLACK FOREST OF LIGNIN (LARGELY UNEXPLORED)
PETROLIA
PLEXIGLAS
NEOPRENE
PETROPOL
CHLOROPRENE (SYNTHETIC RUBBER)
CRYSTAL MOUNTAINS
ACRYLIC
MARATHON
LUMARITH
LUCITE
BAKELITE
ACETIC ACID LAKE
THIOKOL
GREAT ACETYLENE RIVER
CELLULOSE ETHERS
HERCULES
PLASTACELE
SARAN
VINYL
ETHOCEL
CELLULOSE
BUTVAR
CELLULOSE
CELLULOSE ACETATE BUTYRATE
TENITE II
ACETIC RIVER
ACETATE
BUTACITE
HERCOSE C.
FIBESTOS
REGENERATED CELLULOSE
TENITE
CELLOPHANE
SEA
RAYON ISLAND
SYNTHETICA
A NEW CONTINENT OF PLASTICS
N
C
H
O

CAMEL
CAMEL

2

'Miracle is a catchy word…it excites the eye and titillates the imagination. It was applied, indiscriminately and profusely, to nylon… While the Du Pont Company…officially persisted in abstaining from applying the word to nylon, virtually everybody else did – and still does. The lady who wrote: "Dear Du Ponts: I have read about the wonderful new stockings made from dirty old coal – surely you must have the angels working for you" has lots of company.'

25th Anniversary of Du Pont Nylon, 1939–64

1930–50

'N' Day: The Dawn of Nylon

On 27 October 1938, Charles Stine, the Du Pont Company Vice-President, announced to the *New York Herald Tribune*'s World of Tomorrow Forum at the World's Fair preview 'a new word and a new material: Nylon'. This, he told his 4,000-strong, predominantly female audience, was 'the first man-made textile fibre derived from coal, water and air'. Rapturous applause broke out when he described the filaments as being 'strong as steel, yet fine as a spider's web'. These fibres would make the indestructible stockings of the near future. Three days later, the first advertisement linking nylon to hosiery appeared in the *New York Herald Tribune* which also carried a long editorial on the 'fine, shiny, strong' fibre that was 'to be used primarily in the manufacture of hosiery and knitting'. Many other newspapers covered the announcement of this marvel but few could possibly have appreciated just how historic was Dr Stine's speech; it proved to be the opening shot signalling the huge textile and fashion revolution that nylon was to ignite.

Just four weeks earlier, the *Science News Letter* had broken the story about Du Pont's newly patented 'artificial silkworm', describing a viscous fluid which could be drawn into fibres that are 'finer than natural silk and yet have amazing elasticity'. Tomorrow's 'silk' stockings were to be fashioned by chemists – from basic 'caster oil and coal'.[1]

'Du Pont's New Fiber More Revolutionary Than Martian Attack' claimed an impressed *Philadelphia Record* on 10 November 1938. It was the 'Martians of Delaware' who were actually 'remaking our planet'.

Opposite: A cigarette girl selling nylon hose at Philadelphia's Latin Casino. *Better Living* magazine, 1950.

Above: In 1931, Wilmington's *Evening Journal* broke the news that a silk-like fabric could be made by combining antifreeze and castor oil. Opposite: Dr Wallace Carothers headed a small group of scientists at Du Pont's Chemical Experimental Station where they unravelled nature's secrets.

Making play on Orson Welles' famous radio dramatization of H. G. Wells' *War of the Worlds* (which enacted an extra-terrestrial attack so effectively that it had just caused havoc across America), it described Du Pont's own 'chemical coup d'état':

'While panic-stricken motorists were whizzing past the New Jersey police, crying, "Run for your lives! The Martians are coming!" and while men huddled in basements with loaded shotguns waiting for the poison gas of the imaginary messengers from Mars, the world had, for two days, been in possession of information far more startling and in many ways more disturbing...Du Pont research chemists had perfected a new synthetic material named 'nylon'... capable of hundreds of uses, including all types of delicate textiles, brush bristles, racquet strings, fishing lines, dress goods, velvets, knitted and woven underwear and plastic compositions.'

The implications of this discovery were 'so incredible that they could barely be summarized', ushering in the death blow to the oriental silk trade and likely to cause more strategic damage than would the sinking of Emperor Hirohito's navy. In rural America, news of nylon was received with optimism in the coal fields and alarm on the cotton farms. It drove another nail into the coffin of the ailing American cotton industry and shifted the centre of fine textiles from the tropics to the temperate zone.

Charles Stine was the instigator of this earth-shaking discovery. Like Tetley at Courtaulds, Stine was a one-man powerhouse, with a driving vision all his own. A man who 'with his own hands' had manufactured the first TNT for the Navy, he had 20 years' experience in developing Du Pont's synthetic dyes and explosives. Appointed Chemical Director in 1924, only four years later he was able to persuade the board to set aside $250,000 for speculative 'pure' or 'fundamental' research, as compared to research with definite profit-making potential. Ironically, this 'patient money', as it was called, quite unexpectedly spawned a mammoth synthetic-fibre industry that created both a fast and a vast fortune for Du Pont.

Molecular Architects at Purity Hall

None of these developments were anticipated when Charles Stine first set up his Pure Science Division in a new laboratory jokingly dubbed Purity Hall. In 1927, Stine finally persuaded the brilliant experimental scientist Dr Wallace Hume Carothers to leave his teaching post at Harvard to lead the Du Pont research team. In a letter to a friend shortly after his arrival, Carothers described the remarkably generous level of support given to him alongside the 'apparent corporate indifference' to research goals: 'As for funds, the sky is the limit. I can spend as much as I please. Nobody asks any questions as to how I am spending my time or what my plans are for the future. Apparently, it is all up to me.'[2]

Carothers was now in a perfect position to pursue his interest in the controversial theories of the German organic chemist, Hermann Staudinger. In 1920, Staudinger had outlined the basic theory of polymer chemistry by describing polymers as being composed of molecular chains of practically limitless length and molecular weight, held together like the beads of a necklace. Carothers set out to prove scientifically the validity of Staudinger's long-chained molecule theory, and with the help of Julian W. Hill, an MIT graduate, he succeeded. Although the invention was based on his work, Carothers was not actually in the laboratory when the first nylon fibre was pulled from a beaker by Julian Hill. The story has it that so much excitement was generated by nylon that scientists ran about the laboratories to see how far they could stretch the fibre, covering the space like a giant web. The first viable synthetic fibre had been 'drawn' in 1934 and, from the five possible solutions, it was polymer 6–6 that yielded the most viable nylon fibre in 1935. (It was known as 6–6 because each of its constituent molecules, hexamethylene-diamine and adipic acid, had six carbon atoms.)

Fortune magazine described Dr Carothers as 'a brilliant artist in the manipulation of invisible worlds'[3] and his original work was to become a source of inspiration for chemical researchers everywhere. Carothers' earliest experiments established the principles for making polyester fibres (polymers of recurring carbon and oxygen molecules), but he set these aside, being more interested in the promise of polyamides (polymers of carbon, oxygen, nitrogen and hydrogen molecules, later named nylon).

Coal into Silk

The Du Pont board became increasingly optimistic about a tangible and profitable outcome from Carothers' research. In preparation for this, Du Pont filed a broad patent in 1930 covering most of the possible product applications of Carothers' work on 'linear condensation polymers…of infinite length' and described production of 'pliable, strong, and elastic' fibres which could be used for 'artificial silk, artificial hair bristles, threads, filaments, yarns, strips, films, bands and the like'. A year later, the *New York Times* announced that Du Pont was developing a fibre 'at least as good as silk' from a blend of antifreeze and castor oil. Some scoffed at this prospect, such as the comment in the *Detroit News*: 'Man, after experimenting for years, has finally discovered that by an ingenious mixture of castor-oil, ethylene glycol, carbon, hydrogen, and oxygen, he can make a silk fiber almost as good, and not more than three times as expensive, as the one a Chinese worm has been manufacturing for centuries.'[4]

Science News Letter took great pains to distinguish the new silk from rayon. It was not 'made from the cellulose of growing plants like cotton or wood but from coal tar derivatives'. Coal had already produced thousands of useful compounds, including 'perfumes which nature never knew' alongside explosives and dyes.[5] Nylon, in fact, contains the same chemical elements as silk – carbon, oxygen, nitrogen and hydrogen – which are, likewise, the chemical elements of coal, water and air, but the processes by which bituminous coal was transformed into nylon hosiery were many and intricate. *Textile World* gave the following clear description of nylon's chemistry:

'Nylon is made by heating the proper intermediates – a dibasic acid and a diamine, for example – at a temperature somewhere between 400° and 600° F [200° and 315° C]. When that is done – and here you have the secret of Nylon – the molecules begin to hook themselves together in long chains. Nylon is first formed in icy-white ribbons of any width and thickness that happens to be convenient for handling in the factory. If intended for use as a textile fiber, these ribbons are broken into little chips. The chips are melted, and the water-clear liquid, looking like thick glycerine, is squirted through tiny nozzles to form cobwebby filaments, which solidify in the air and are wound on spools…stretching makes Nylon stronger.'[6]

The making of nylon fibres was not simply a matter of chemistry; every piece of equipment needed to

complete the process also had to be invented by the Du Pont researchers. Nylon was first spun in 1934 in a small, squat, windowless hut formerly used to store explosive powder, and one of the research scientists then involved recalled the almost garden-shed improvisations that made the first fibre possible: a spinning machine was made 'using five inches of second-hand brass pipe and a hypodermic needle'.[7] There were also problems in producing enough of the polymer. The first small chunks of nylon were picked over for hours in order to remove glass segments from the test tubes, then the nylon piece was wrapped in a towel and hammered to break it up, and then the chips were picked over again to remove the towel linters. Many precious years were spent in 'luckless experimentation' until, in 1936, a ray of hope appeared when it became possible to spin for 'as long as ten minutes'.[8] It was only when more scientists and technical experts visited their laboratory that the researchers felt that they were 'really going places'.

The apparent suddenness of nylon's arrival as reported in the press was an illusion – it had in fact taken almost 11 years of research, cost $27 million, and involved more than 230 scientists and technicians.[9] Stine's research programme, which had begun in 1928, had resulted in the super-polymer first synthesized in 1935. It is difficult to determine at which point the vast financial potential of Dr Carothers' work began to be understood – and he himself did not live to witness the consequences of his original research. He never heard the word nylon, nor saw a pair of nylon stockings, nor witnessed the huge commercial success of polyester and acrylic fibres. For many years Carothers had suffered from bouts of instability and depression and in June 1936 he suffered a nervous breakdown from which he never fully recovered. On 29 April 1937, just 20 days after the patent application for nylon had been filed and four months after his marriage, the 41-year-old scientist booked into a Philadelphia hotel and finally took the cyanide that he had carried for many years. After his suicide, a posthumous patent (No 2,130,948) was granted by the U.S. Patent Office and the *Science News Letter* wrote that 'this strange fibre promises to be silk's crucial rival in the hosiery field'. This was a gigantic understatement: as became increasingly obvious, nylon was to prove momentous in every sense. It was of the moment; exactly the right fibre for the right purpose at the right time.

A Tarnished Image and Nationalistic Zeal

During the early thirties, a unique set of circumstances existed which was to guarantee the success of Carothers' new polymer fibre. First, the Du Pont Company was suffering from an image crisis and needed to project itself in a new guise. Second, there was an urgent desire to make America self-sufficient in raw materials by developing man-made alternatives to goods and materials it had to import.

Du Pont had emerged from the First World War with, quite literally, an embarrassment of riches. Described as the 'first chemical war', it would have been difficult for Du Pont to avoid a profit boom, but the family came under personal attack in 1934, accused of wartime profiteering. In *American Plastic* Jeffrey Meikle describes how Lammot Du Pont and his brothers faced a three-day grilling from a U.S. Senate committee which decided that the wartime leap in their annual profits from $5 million to $60 million was gained by overcharging the military. But as Jeffrey Meikle points out, more than 90 per cent of Du Pont's output was for 'peace-time products'.[10] A satirical cartoon that appeared in *Forum* magazine in July 1934 marked the decisive turning point for Du Pont, the public jibe that prompted a serious change of image. Under the heading 'merchants of death' it carried an 'endlessly repeating list of munitions manufacturers, from Du Pont to Krupp and Mitsui', against which were listed their products: machine guns, poison gas, dirigibles, gunpowder etc.[11] In the aftermath of a world war, the manufacture of explosives was a socially unacceptable way to make profits and (as far as popular myth was concerned) Du Pont were '*the* powder people'. It was a matter of urgency to shake off their unsavoury association with killing commodities and to acquire a more consumer-friendly image.

In the face of this open criticism, the Du Pont board snapped into action by appointing the equivalent of today's 'spin doctor' – Bruce Barton, of Batten, Barton, Durstine & Osborn, the most famous advertising man in America at the time and author of a controversial advertising book portraying Jesus as a 'consummate business executive'.[12] Barton advised Du Pont to spend $650,000 on an immediate image reversal so that they would be seen to be designing for living rather than for obliteration. BBD&O came up with a now famous slogan which would express Du Pont's future domestic role: 'Better Things for Better Living.... Through Chemistry'. First

Du Pont produced 1,100 different nylons by the late 1940s. These differed in denier size, twist, brightness or number of filaments.

introduced in 1935, this important message was relayed across America in a series of sponsored radio programmes, *The Cavalcade of America*. Patriotic in tone, the programmes were punctuated by news of the life-enhancing wonders invented by Du Pont chemists.[13] Barton's vision was to associate Du Pont with 'style' in the mind of the public, a bold move, as, at this time, rayon was one of Du Pont's few consumer products but they had had limited success selling it.

This was motive enough for the Du Pont board to take an interest in the fibre discoveries then emerging from Stine's laboratories, but there were other and equally pressing grounds for their desire to discover a substitute silk fibre. America and Japan were in economic and political conflict and at the epicentre of this strife lay, incredibly, ladies' hosiery. Japan was the source of 90 per cent of America's raw silk imports, worth $100 million in 1938, three-quarters of which were used for stocking making[14] (viscose was considered to be too 'lustrous, inelastic and insufficiently sheer' for the American market). The daily purchase of silk stockings in the United States was 1.55 million pairs, representing an annual expenditure of $475 million. It is ironic that something so flimsy, so feminine and so fragile could fire such international antagonism. The United States government, the stocking makers and American citizens themselves resented the fact that they were almost completely dependent on Japan for their most prized luxury fibre.

The First World War had taught Americans that they were vulnerable to supply famines of many vital materials which, by necessity, had to be imported. Germany excelled in organic chemistry, in dye-making, pharmaceuticals and the 'hundred other necessities, which for many years had been their industrial monopoly'. Iodine and rubber, too, were imported, and nitrates – essential for making explosives, fertilizers and fibres – were brought in from Chile. Self-sufficiency was a matter of survival for the American people and for American industry, which Du Pont clearly recognized: 'Until this country could build up its own resources, its needs could be met only at the whim of a foreign power.... Such things were irritating in peace, for they forced the country to pay a sort of ransom to attain the living standard it desired. And in war this dependence on foreign technology could well prove fatal'.[15]

Seen in this context, the press excitement surrounding Stine's announcement is understandable. The Du Pont archives have an extensive collection of press cuttings from this period, many of which reflect the euphoria that welcomed their synthetic fibre as being a triumph over Japan: a *New York Times* headline says '$10,000,000 Plant to Make Synthetic Yarn; Major Blow to Japan's Silk Trade Seen';[16] 'Silk Worm Turns Green' (with envy) from the *San Francisco Dividend Advocate*; 'New Synthetic Fiber May Smash Jap Monopoly', proclaimed the *Philadelphia Record*. Japan's invasion of Manchuria and attack on Nanking had already alienated millions of Americans and one irate young woman wrote to Du Pont to say that she 'did not intend to pay for a single silk stocking until the Japanese get out of China'.[17] Silk boycotts had been tried but without any real success so that the *New York World Telegram* regarded Du Pont's synthetic silk as 'having the utmost social and economic significance... It won't be so difficult to popularize a boycott of Japanese silk when women can obtain stockings from the Du Ponts' mechanical silkworm that are not only equally attractive but wear longer.'[18]

Don Wharton's report in *Textile World* (1938) pointed out that the United States bought four-fifths of all the silk used in the world and four-fifths of what they imported was knitted into stockings. He went on to say that although the 4,000,000 lb (1,800,000 kg) or so of nylon fibre estimated to be made in Du Pont's first year of production would 'not immediately take over a market which last year used 45,000,000 lb [20,250,000 kg] of silk', the Japanese were 'watching anxiously'.[19] And they were: the Japanese silk producers made representations to their government to do something about the threat of nylon stockings.

The 'wildly fluctuating silk market' had the American hosiery makers 'against the wall.... confused and confounded by a product that might cost $1.50 a pound one day and, without any apparent reason, $3 a week later, [they] turned eagerly to nylon. It presaged a stability in supply, quality and price that would allow them to operate their small businesses on a rational basis.'[20]

Trade protectionism was a hot issue in the United States during the depression-blighted thirties and attention began to focus on The American Viscose Corporation (AVC) which was the world's largest rayon producer and one of Courtaulds' many, and highly profitable, subsidiaries. In July 1937, *Fortune* magazine was telling the American public of the unknown dividends that were being salted away by the British. 'American Viscose, modest, secretive, and unknown, is one of the industrial miracles of our time'. A comparable phenomenon, it told its readers, 'to Standard Oil or the automobile empire of Henry Ford'.[21] It was economically unpalatable for such an important textile business to be in the hands of a British company and in March 1941 the American newspapers carried the headline 'Bang Goes Viscose', describing the enforced sell-off of the British AVC to an American bank syndicate for $100 million. This was a great loss to Courtaulds, effectively stripping it of about half its viscose business.

True enough today, but will the roles be reversed five years from now?

An Identity for the Bright New Star

American industry was shaping up to make the most of its new synthetic fibres. Du Pont had spent millions of dollars before the first nylon was produced but, as the potential began to be recognized, the company started to invest heavily in new plant for its manufacture. The scale of the whole Du Pont enterprise was summed up in 1940 by *Fortune*:

'The U.S.... is witnessing the premiere of a new industry that some day may prove even more spectacular than the $252 million rayon-yarn business. There is, by way of stage setting, the new, streamlined, well-guarded nylon plant that rose out of a cow pasture and into production last December before the startled eyes of the truck farmers and fruit-basket makers of sleepy Seaford, Delaware – tangible expression of an $8,600,000 investment. There is the addition of a nylon division to the line of basic chemicals, rayons, plastics, dyestuffs, paints, finishes, explosives, cellophanes, and synthetic rubber that together make $858,000,000-Du Pont one of the great chemical syntheses of the world. There is nylon's visible invasion of the $400,000,000-a-year full-fashioned silk-hosiery market. And there is the fact that nylon may easily gross some $11,000,000 for Du Pont in its first year.'[22]

Before Du Pont's fibre could be launched it had to be given its own identity, and not the least of the problems was to find an appropriate name for the new product. 'Certainly women could not be expected to step up to a hosiery counter and blithely order a pair of hexadecamethylene dicarboxylics', as one droll reporter suggested after Carothers and Hill made their original report.[23] The working name

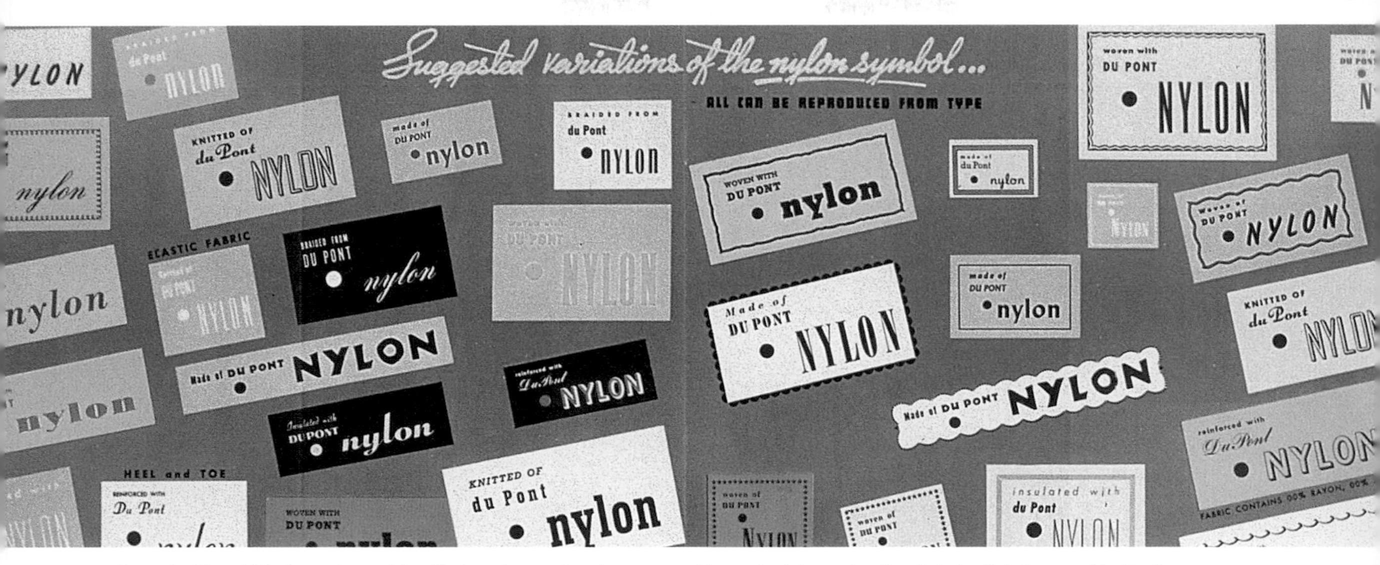

Opposite: To satisfy its huge demand for silk, America was largely dependent on Japanese imports, but by the early 1940s, Du Pont's 'nylon' was already causing pain to Japan's mighty silk industry.

Above: A rainbow of options in 'nylon' labels created by Du Pont for its clients' use.

'Rayon 6–6' became 'Fiber 66' in 1937.
The company's top executives had by then realized that it was crucial to come up with just the right sounding name and identity for their new fibre – one that would distinguish it from rayon and, above all, break with the notion of a synthetic fibre being an 'artificial' substitute. The Name for Fiber 66 Committee was convened and a Coordinator of Suggestions appointed. Around four hundred ideas were put forward, including Amidarn, Artex, Dusilk, Dulon, Linex, Lastica, Morsheen, Novasilk, Nurayon, Nusilk, Ramex, Silpon, Self, Tensheer and Terikon.

Dr Ernest K. Gladding, later to become manager of Du Pont's Nylon Division, who, along with Elmer Bolton, had the inspired idea to perfect Fiber 66 for use in women's hosiery and to launch it in competition with the luxurious fully-fashioned stockings at the top end of the market, was at the final marathon meeting which decided on the name nylon. His own personal favourite was 'Duparooh' – a playful abbreviation and acronym of 'Du Pont Pulls A Rabbit Out Of Hat'.[24] There were many possible names on the way to 'nylon': 'Gladding zeroed in on "norun". It was catchy and to the point. But in the light of tests showing that Fiber 66 *did* run, to adopt a name that promised otherwise would have been a public-relations fiasco. In mounting frustration, Gladding tried flipping "norun" into "nuron". This had the drawback of sounding too much "like a nerve tonic", given its unfortunate resemblance to "neuron". Not to mention "moron".'[25]

One myth that frequently surfaces is that 'nylon' was derived from the combination of New York (NY) and London (LON): incorrectly believed to be the twin sites of nylon's simultaneous discovery.
To summarize Gladding's account of the derivation: from NuRon, they changed the 'u' to an 'i', but that could be pronounced in several ways which would have been confusing. The 'i' was changed to 'y' and, after trying various alternatives, Nylon was adopted to describe the base material 66 in early October 1938 – just two weeks before its public launch.[26]

The word nylon was introduced to the world as the generic term for polyamide fibre but it was never patented as a trademark by Du Pont. It went into the international language as a free word and the following is Du Pont's own definition: 'Nylon is the generic name for all materials defined scientifically as synthetic fiber-forming polymeric amides having a protein-like chemical structure; derivable from coal, air and water, or other substances, and characterized by extreme toughness and strength and the peculiar ability to be formed into fibers and into various shapes such as bristles, sheets, etc.'[27] Nylon was to be a generic word describing a material, much like 'steel' or 'wood', and covering a whole family of potential textiles and other products, characterized by being lightweight, strong, elastic, non-absorbent, non-deformable, non-shrink or 'self-ironing'; it was even used as an enamel.

Du Pont's legal department took a considerable amount of trouble to ensure that no other product with the same name was in existence – their 1941 archives show that ICI drew their attention to a 1906 advertisement in the British *Strand* magazine for a disinfectant trademarked 'Nylon'. Much to everyone's relief, the trademark proved to be unregistered and the firm itself no longer existed.[28] Another issue arose when Du Pont's competitors, including Courtaulds, insisted that nylon be categorized as just another rayon – hoping, perhaps, that the 'wonder fibre' boom would evaporate if the new materials were indistinguishable from the familiar generic rayons. However, Du Pont found an ally in ICI: 'You can, without question, rest assured that ICI's views on this matter coincide with those of your company and that they will do everything in their power to ensure that nylon retains its individual identity and is referred to solely by its correct name.'[29] Du Pont won the day. The alarm of British fibre manufacturers was well founded as it was abundantly clear from nylon's brief but spectacular pre-war launch that the switch to synthetics would shatter both traditional textile economies and even undermine the age-old hierarchies of fibre status. It was the context in which the new 'miracle' was first introduced to the public that gave it such credibility and value.

The Wonder World of Chemistry

Once the new fibre had been perfected and named, it was time to develop a programme that would cut out the delay 'between test tube and counter'. Two factors would smooth the way: Du Pont's excellent global public relations department and a tried and tested method of humanizing the image of their chemical products. The latter was through the Wonder World of Chemistry show, first presented at the Texas Centennial Exposition in 1936. This was intended to demonstrate the boon to farmers that was the chemical industry. Anxious to emphasize a healthy connection with nature, the Du Pont promotion centred on the 'partnership between farming and chemistry' and pointed up the agricultural raw materials that were converted into chemical products: cotton, wood, turpentine, molasses and vegetable oils became fabrics, plastics, synthetic rubber, antifreeze, cleaning materials, insecticides, dyes and paints.

The 1.5 million visitors to the Wonder World of Chemistry were told that nature was being 'improved upon' by Du Pont's chemists. This was a significant event, marking the beginning of a new era in visual communication – it would no longer be enough just to manufacture a highly technical product; in future, these unprecedented materials would need to be 'explained' to the consumer and their invisible qualities demonstrated. Chemical manufacturers became the direct link between the laboratory and the consumer. In addition, their role in fibre development became increasingly complex. They bought patent rights, developed products, made prototype textiles and garments and then sold the concept of the new material to commercial textile manufacturers and consumers through exhibitions and trade fairs. Thus, corporations became educators of their trade clients and simultaneously of the fashion consumer.

Two days after Dr Stine's announcement of the birth of the first truly synthetic fibre came the first advertisement using nylon, as bristles in Dr West's Miracle Tuft Toothbrushes, replacing hogs' hair. Du Pont's newest miracle fibre, identified only as 'Exton', was guaranteed not to shed, break, pull out in your mouth, go soggy and limp when wet, or otherwise pose the various and sundry animal bristle troubles. Exton was actually a cruder version of nylon and samples of Dr West's toothbrushes were shown at The Wonder World of Chemistry, demonstrating that they would last two or three times longer than natural bristles. Nylon was also made into fishing lines and surgical sutures but its real impact on both the economy and the culture of the United States was to be in the form of glamorous nylon stockings.

The Miracle of Nylons and the Coming World of Man-mades

Experiments to make prototype fully-fashioned nylon hose began at Du Pont in 1938. Starting from scratch with their new fibre, Du Pont technicians had to solve all kinds of problems before stockings could be produced. First, the fibre had to be extruded then wound onto bobbins, a finish had to be added, and finally it had to be knitted into stockings on machines designed to knit silk. There were many difficulties en route, new machinery had to be made to spin the fibre strands, new coatings developed for yarns and new stitch techniques and knitting constructions devised that would compensate for the physical differences between nylon and its predecessors, silk and rayon.

Cheap stockings were knitted in a continuous tube and often in rayon fibres which bagged and drooped at the knees and ankles, whereas silk hose were knitted flat and were shaped to the leg by dropping stitches and joining the tube with a seam at the back. Although they never manufactured nylon stockings for sale, Du Pont's scientists had to prove the feasibility of this use and the operations along the way included 'melt spinning, pretwisting, draw twisting for strength, uptwisting for elasticity, shrinkage removal, sizing to protect filaments during further processing, spooling, knitting, preboarding of stockings to prevent wrinkles, dyeing, and so on'.[30] Being thermoplastic by nature (changing shape when exposed to heat), the first nylons emerged from the vats as 'a shapeless, shrunken mass'. This setback was turned to an advantage when Du Pont devised the technique of preboarding nylons, a patented method that came to be used throughout the hosiery industry. Under this system, newly knitted stockings were placed on leg-shaped forms and steam was passed through to heat-set the shape and size.

'Nylon Irons Itself' claimed a press release, and the excitement surrounding the launch of the new stockings centred on their sheerness, their smooth wrinkle-free fit and durability. This enthusiasm was just as well as nylon hosiery yarn was introduced at $4.27 a pound (450 g) as compared to silk at $2.79.

Above: Models demonstrated the strength of nylon stockings at the New York World's Fair in 1939. By 1949 some 85 per cent of stockings were made of nylon.
Opposite: Du Pont's Tower of Research housed its Wonder World of Chemistry exhibition which symbolized the triumph of the coming chemical future. It is shown here at the New York World's Fair in 1939.

Although the hosiery potential of nylon had been announced in October 1938 and the first experimental stockings were knitted in February 1939, stockings were not available to the public at large until May 1940. By stimulating high levels of interest over an extended period, Du Pont generated a ready-made demand before nylon stockings even went into production. After a decade in development, the timing and circumstances of nylon's release was brilliantly stage-managed. First seen at the San Francisco World's Fair in February 1939, nylon stockings made their most sensational impact at the New York World's Fair later that year. Costing more than $155 million to build, the New York World's Fair was intended to 'win popular understanding of new product developments'. *Du Pont Magazine* noted that this 'Drama of Opportunity' showed that 'Almost a quarter of all manufactured goods produced in America today are chemical goods, and few articles among the remainder escape the chemist's influence. [Chemical science] is dedicated to change – *to change through creation of new substances as distinguished from change induced by mechanical invention*. Get that fact and what it connotes and you will have mastered the biggest lesson to be learned at the Fair.'[31]

Symbolizing the triumph of modern chemistry was Du Pont's 32 metre (107 foot)-tall Tower of Research housing the Wonder World of Chemistry. Its steel skeleton had 'been designed to reflect the skeletal structural character of carbon-based molecules', the fundamental basis of organic chemistry. Inside was an 18 metre (60 foot)-long mural designed by the artist Domenico Mortellito depicting 'chemistry's part in creating products for "better living" from the raw materials of forest, farm and mine'. Symbolic figures were carved in intaglio relief from varicoloured 'Lucite'[32], a methyl methacrylate resin, and 'Plastacele', a cellulose acetate plastic. This work, the artist believed, symbolized 'a new day in art when new materials of chemical origin will displace the traditional media of expression used by artists for centuries'.[33] As art history has shown, this proved to be a prophetic statement – within 20 years acrylic paints, PVCs and mouldable plastics did indeed become the new media of 1960s Pop art.

Above the main reception desk at the Wonder World of Chemistry were inscribed the words 'Dedicated to the men and women who, through their contributions of labor and enterprise capital to the chemical industry, are creating Better Things For Better Living'. At the exhibition preview in New York, the President of Du Pont, Lammot du Pont, set the tone with a speech stressing the importance of chemical invention to the American sense of progress and even to its national security and political stability.[34]

Everywhere, scientists were portrayed as the new national heroes, pushing America's frontiers further and further forwards, inspiring a new generation with the magic of chemical creation. An article in the *National Geographic* in 1939 drew attention to the huge popularity of the children's Chemistry Set. 'Toy chemical sets, played with by fascinated boys and girls, make smoke, smells, and bubbling magic in a million homes. In countless classrooms other groups stain their clothes and fingers monkeying with test tubes and tiny bottles. No branch of science holds more devotees, or more portent for tomorrow.'[35] In the laboratories of skilled professionals, chemical engineering was making 'one great factory of the whole United States', remarked Du Pont's Dr Stine, and 'membership of the American Chemical Society was 22,245, linked with 2,000 research laboratories – no other country made more new things by synthetic chemistry'. Since 1918, he estimated, more than 200,000 products 'entirely new to man' had come from American laboratories and one third of factory production on the eve of the Second World War were things unknown in 1880.

Du Pont was dedicated to creating and promoting more and more new products, a policy described in 1940 by *Fortune* as 'forcing' nylon into being: 'It was developed under deliberate pressure, in line with Du Pont policy. If no new or improved products are forthcoming at reasonable intervals, then they must be made to appear. This isn't unique in the fast-moving chemical industry, but Du Pont carries it to the length of a religion.'[36] The piece in the *National Geographic* extols the virtues of chemistry: 'Man needs the chemist to give him precious things'. His women want to wear amber, jade, ivory, crystal, even pearls and other beautiful stones, and these man can't always afford because they are so rare and dear but now, by synthesis, chemists make them at low cost.... even the rare perfumes which women use to 'attract the male' can be imitated with coal tar musk.... Some things chemists do now would have been classed in Bible times with the miracles.... As when water was turned into wine.' Chemistry has given us 'the magic of soft panties from lumps of coal, sugar from pine trees, or delicate heliotrope scent taken from stinking tar', all akin to the 'wizardry of ancient alchemy.' Nature is no longer 'the boss'.[37] In keeping with this utopian period in synthetics' history, man was regarded as being in some sort of contest with nature, and science, by unlocking nature's secrets and emulating its materials, would lead humanity towards a secure and triumphant nature-free future.

Du Pont's image was totally metamorphosed in the public's perception after the introduction of nylon – no longer behind munitions, instead it was a company behind lovely legs. Nylon was 'a textile prodigy' and received by the people like a conquering hero. It became a household word in less than a year and, in all the history of textiles, no other product has

Opposite: Fixing hose into permanent leg forms. Nylons were preboarded and heat-set using a specially patented Du Pont method.

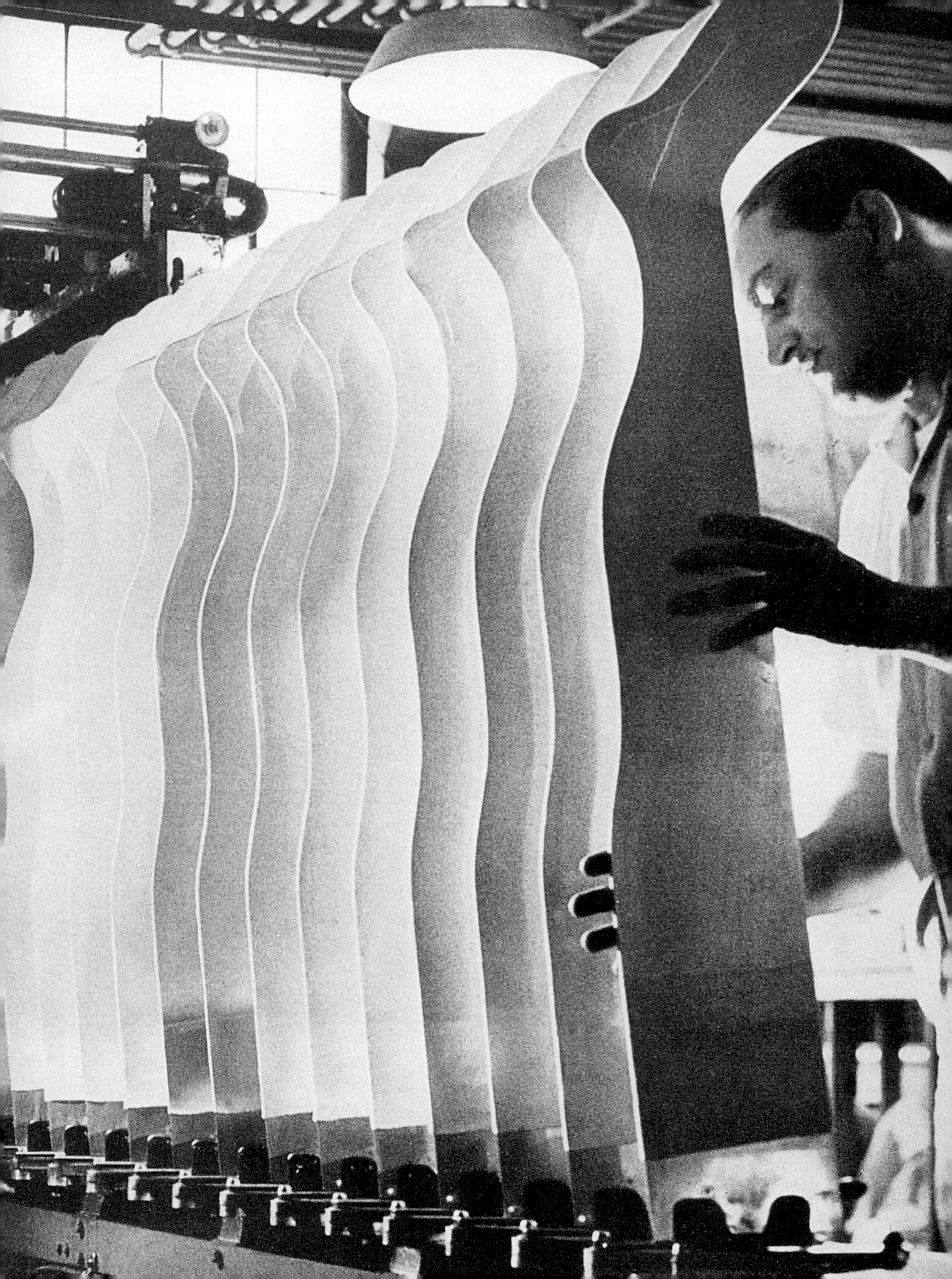

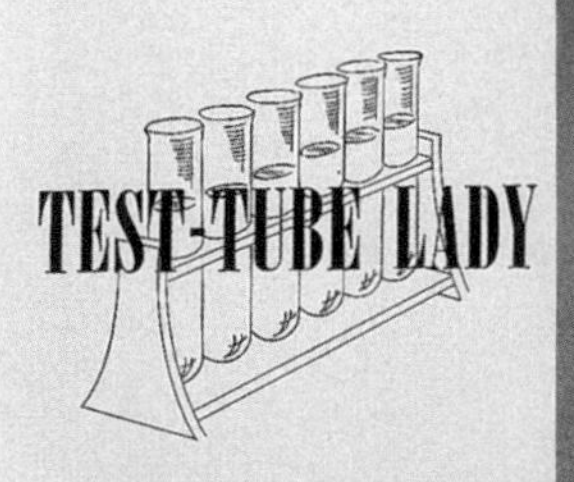
TEST-TUBE LADY

FROM the tip of her heels to the topmost curve of her new hat . . . and fashionably speaking only, mind you . . . the modern woman is more and more a product of the plastics laboratory.
Here, for instance, is "Princess Plastics."
Her bright, sleek heels carry "Pyraheel," Du Pont plastic heel covering, which will match almost any leather effect and still resist scuffs.
The trim of her hat is "Plastacele," in this case glistening black, belying the whole range of bright colors which come to life in this Du Pont plastic.
Plastics, too, are her slide fasteners, shoe bows, lipstick container, and even her crystal-clear walking stick.
But notice, especially, her necklace and bracelet—entirely new pearl effects in pink and white, made possible with "Lucite." And her sapphire cosmetic carry-all represents a clever molding job with "Lucite" molding powder.
Whether she realizes it or not, Princess Plastics and thousands like her represent a real market for plastic products. When you're designing for the style field, make sure you have the varied qualities of a whole group of versatile plastics at your command. Write Du Pont, Plastics Dept., Arlington, New Jersey.
"Lucite" methyl methacrylate resin —molding powder, sheets, rods and tubes
"Plastacele" cellulose acetate —molding powder, sheets, rods and tubes
"Pyralin" cellulose nitrate —sheets, rods and tubes
"Butacite" polyvinyl acetal resin —sheets
DU PONT PLASTICS

THE DU PONT MAGAZINE
JUNE 1940
THE CHEMICAL GIRL OF 1940

enjoyed the immediate, overwhelming public acceptance of Du Pont nylon. 'Before the year is out, millions of American women will be wearing hosiery made from coal, water and air' commented *Textile World* in 1940. Du Pont were determined to avoid the fate of rayon, which became branded as a cheap substitute for silk by being made into products for which it was ill suited. Highlighting the spirit of honesty and pride with which nylon garments were to be produced, Du Pont pointed out that 'The word "rayon" wasn't put on all rayon products until the law compelled it. The word "nylon" will be plainly stamped on or knitted into every pair of nylon hosiery'.[38] Depending on the gage, prices for nylons were expected to be $1.15, $1.25 and $1.35 a pair, 'corresponding to the prices of fine silk stockings so that nylon will be established as a superior product'.[39]

Du Pont went to elaborate lengths to keep their new fibre secret before its launch, and were similarly cautious about guarding their star product, stockings, in the months leading up to its public release. The first stockings were sold to female employees of Du Pont on 20 February 1939. Restricted to two pairs of stockings apiece, in a single shade and costing $1.15 each, the girls were asked to complete a questionnaire on the stockings within 10 days of purchase. The first public sale was in stores in Wilmington, Delaware on 24 October 1939. Although customers were limited to three pairs and only locals were eligible to purchase, 4,000 pairs of stockings were sold in just three hours.

It was one of those occasions when technology and fashion met in perfect unison. Shorter skirts threw attention on to the legs and, of the existing stocking options, silk was expensive and prone to runs, rayon was badly fitting, with a give-away highly lustrous shine that shouted 'second-rate', and cotton (lisle stockings) were thick and sagged around the ankles and knees. By contrast, based on the statement that 'nylon had the strength of steel and the sheerness of cobwebs', the press and public assumed that nylon stockings would not run, and one humorist warned women that they would have to 'remove their stockings with an acetylene torch'.[40] Du Pont had overplayed its 'no more runs' promise and was uncomfortable with the 'miracle yarn' image of nylon. It was vital that they tempered public anticipation with a campaign urging women not to expect the impossible. A press release was issued in order to 'disarm the ridiculous rumours that the stockings are runproof, that two pairs will last a woman a year, that the threads are impervious to razor blade and nail file, and that you can't burn a hole in one with a cigarette...they are not runproof.... Nylon hosiery is susceptible to minor snags, but the punctures are less likely to develop into runs. Some hypersensitive women report that Nylon hose have a slightly colder feeling than silk stockings and lack silk's clinging touch.'[41]

Synthetic Luxury and Superior Science

By the opening of the New York World's Fair in 1940, Du Pont was ready to show off its new nylon material at the Wonder World of Chemistry show with even greater confidence than in 1938. It was a dramatic display in a suitably dramatic setting.

'In a floodlit courtyard, an illuminated fountain composed of thousands of nylon filaments, animated by air currents and spotlighted at night, set the stage for the world premiere of *Nylon: This Chemical Marvel*. A knitting machine knit nylon stockings right before your very eyes. A pair of mechanical hands stretched a nylon stocking to its longest length and snapped it back, day and night, around the clock, proving to sceptical visitors that nylon was surely the strongest, most elastic fiber ever invented.... Du Pont's floor show – better known as a "leg show" – was judged "sexiest corporate show at the fair" by one impassioned observer.'[42]

Statistics were displayed to show that Du Pont were investing $7 million a year in research that would produce 'Better Things for Better Living' and a very significant section of their huge display was entitled 'Chemistry Serves the Realm of Fashion'. Du Pont brought nylon to life and fashion to nylon by parading prototype clothes and accessories on beautiful models – with constant attention on the girls' legs – before an awe-struck and appreciative audience. The masculine world of molecular chemistry was, for the first time, portrayed as leading fashionable women forwards. These expensively produced prototype designs were essential to the work of the Style News Service offered by the Public Relations offices that Du Pont maintained in New York's Empire State Building. From here, bulletins and photographs would be circulated all over the world, based on the principle that the only way to stimulate the interest of editors and manufacturers alike was to show examples of real,

Above: The 'living symbol of the Chemical Age', illustrated in the *National Geographic* in 1939. Her complete outfit, 'all of laboratory origin', is made from cellophane, rayon, plastic and 'Lucite'. Opposite, top: Princess Plastics, wearing an ensemble made from 'Plastacele' cellulose acetate, 'Pyralin' cellulose nitrate, 'Butacite' polyvinyl acetate resin sheet and 'Lucite' methyl methacrylate resin. Opposite, bottom: The test tube would provide plentiful fabrics and luxuries of the feminine dressing room, including 'pearls' of 'Lucite' and shoes of synthetic leather, as worn by the Chemical Girl in 1940.

desirable, garments: it was hard to get excited by looking at an image of a fibre.

The Style News Service circulated the story of the heroines of the New York World's Fair, the Test Tube Lady (also known as Princess Plastics), who was shown emerging from a giant test tube and totally decked out in artificial materials, and Miss Chemistry of the Future, who modelled all-nylon 'lace evening gown, stockings, satin slippers, and undergarments'. Tantalizingly, as a Du Pont press release stated, this was decidedly the 'costume of tomorrow' since most of the nylon products shown were not yet commercially available. Mainstream America was enthusiastically informed about the sensational synthetic fashion future ahead in an illustrated article in the November 1939 issue of the *National Geographic*, opening with the following paragraph:

'At a New York fashion show, we saw a girl clad from head to foot in artificial materials. Everything she wore was made from synthetic stuffs created by chemists. Her hat was Cellophane; her frock was Rayon. She wore "Nylon" stockings and carried a patent-leather handbag and stood in imitation alligatc shoes and wore "jade" bracelets and "ivory" beads; her parasol handle was from beautifully colored plastic. Even the faint hint of musk on her imitation silk handkerchief came from a synthetic perfume; on her nails there glistened a synthetic dye, and other coal-tar dyes imparted rich shades to her ensemble.'

The 'startling symbol' of the new world was this woman clothed entirely in man-made materials, fibres, plastics and synthetic dyes and Frederick Simpich, the author of the piece, posed the question: 'What does it all mean? What needs or forces suddenly set all our laboratories at this miracle-making?'. He found his answer in the words of Du Pont's Dr Stine: no longer would sheep, plants and worms provide the fibres for textiles and clothes; accessories would not need to be made from bone, hide, horn, insect excretions or the saps and barks of trees. Until the age of plastics, homes had been built from the same materials for thousands of years – stone, brick and wood – and, worse still, 'we were wearing the same clothing as our great-grandfathers ...merely cut to a new style and woven by machine instead of by hand. We were eating the same foods, using the same perfumes, sleeping in the same types of beds that the Caesars and Pharaohs knew.'[43] Now that a 'wave of inventive genius' had swept over the chemical world, the traditional materials were newly understood to be merely compounds of raw materials that 'are ever present in superabundance in the air, the water, and the soil'. With the first great break in the materials tradition, many, many things were 'born in a laboratory', including 'combs, brushes, buttons, buckles, bracelets, pens, pencils, fingernail tint, costume jewelry, fishing tackle, candy and hat boxes, cigar, cigarette and food wrappings – even your wife's hat, hose, underwear and gowns, as well as your window shades and draperies'. Cellophane was 'another white rabbit from the chemist's hat' which, being extruded through a slit instead of fine holes, came out in flat transparent sheets and was used for clothes and accessories as well as for food wrapping .

In 1940, Du Pont made the modest prediction that 'In the dress-goods field the new Du Pont yarn appears to offer promising possibilities for future development'.[44] At this time, experimental foundation garments, lingerie and knitted fabrics containing nylon began to appear in department stores. Products such as 'All-Nylon Lingerie' made by the Holeproof Hosiery Co, were selling out as fast as manufacturers could get the yarn – despite the fact that prices for 100 per cent nylon tricot lingerie were the same as for top quality pure-silk garments. Probably the most extensive nylon line was that of the Formfit Co. It first offered two 'Nylies' garments: a girdle, to retail at $5, and a combination girdle-brassiere at $7.50 (later, Formfit was very successful with high-priced nylon items in its evocatively named 'Schiaparelli' range). Formfit hit the jackpot with 'Nylies' because they were soft, boneless and appealed to 'the younger trade of slight or average figure.... Nylon lastex outcontrols silk lastex or rayon lastex, say its friends, and has less weight and bulk. Like nylon hose and nylon lingerie, nylon lastex washes easily and dries quickly – a major advantage to the gal who is able to afford just one high-quality foundation.... even more astonishing is the assertion that "nylies" do not acquire the wearer's shape after she has worn them for a few months.'[45]

Opposite: No need to open a cellophane suitcase at customs: an American girl with the latest in luggage at a New York pier, illustrated in *Picture Post*, c.1947.

Above: 'N' Day, 15 May 1940, the first day of the nationwide sale of nylon stockings in the United States. The enthusiasm for nylons led to crowds of shoppers storming department stores. Some 780,000 pairs were sold on the first day.

Nylon's launch in the intimate guise of stockings and lingerie was a perfect example of the successful domestication of a highly technical product. Here was something every woman could understand. The eternal problem with scientific innovation is how to translate it into product and it often takes decades before a new fibre or fabric can win over the clothing manufacturers and finally reach the consumer. One major reason for this is that with every change of raw material comes problems with the machinery that is used to knit, weave, cut or to make up the garments and sometimes a whole change of plant is necessary as well as expensive retraining for staff.[46] Nylon was different, it was a textbook case of how to do it: the fibre developer reached the consumer and the press directly with the promise of a small but luxurious product that was a fashion fixation at the time and the manufacturers had to supply the goods.

Although Du Pont invested more than $4 million in development, they could have had little inkling of the momentous craze that 'nylons' would introduce. 'Nylon wasn't a word in any language two years ago' the July 1940 issue of *Fortune* noted, but since then it had generated a 'flood of rumor and counter-rumor', detonated wild speculation about its effects upon the U.S. economy and 'loosed a stampede' on hosiery counters. These were 'all the sound effects of a revolution'.[47] The whole nation was waiting for the nylon utopia. Nylons became a potent symbol of American technical know-how but they were also redolent of all the escapist glamour of Hollywood. This is not to say that there was not a certain amount of consumer apprehension, the shadow of suspicion and distrust that accompanies all chemically based 'unnatural' inventions. Some critics believed that they would get nylon poisoning, or coal stains on their legs.

'N' Day: 15 May 1940

Mills and department stores went into overdrive preparing for 'N' Day, 15 May 1940, designated as the day when, simultaneously, from Maine to California, nylon stockings would go on sale. It was the day when

Above: A two-ton model of Marie Wilson's leg is unveiled by a Los Angeles hosiery shop. The actress is hoisted skyward for comparison. Opposite: A nylon-hosiery vending machine in a Californian airport. The modern fibre was for modern lifestyles. *Better Living*, 1950.

nylons went national and, in preparation for this and on the eve of the great event, Du Pont made a special nationwide radio broadcast in *The Cavalcade of America*, during which 'questions about nylon stockings now in the minds of thousands of women will be answered by Dr G. P. Hoff, Director of Research of the Nylon Division of E. I. Du Pont de Nemours & Co.' A typical housewife was selected to represent the consumer's point of view and Dr Hoff answered the questions: 'How will nylon stockings wear, what will they cost, who makes them, what care do they require, and what are some of their other characteristics?'[48]

The next morning, crowds queued for hours waiting for department stores to open and their hosiery counters were thronged. Newspapers had a field day: 'Girl Collapses, Woman Loses Girdle at Nylon Sale', 'Battle of Nylons Fought in Chicago', 'Nylon Customers Swamp Counters as Opening Gun is Fired'. When the doors opened crowds surged down the aisles and customers were 'two and three deep' at the hosiery counters.

'Apparently plenty of husbands had been commissioned at breakfast tables not to return home without nylon stockings, for there wasn't a department store in which a good many men did not appear, each holding a little slip with sizes jotted down on it.... Stores sold one pair per customer – sales assistants were told not to part with an extra pair "not even if your grandmother wants it". Larger stores took on...between 10 and 30 girls to cope with the rush.'[49]

In spite of the fact that nylons were priced at the same level as silk stockings, the limited supply of about three-quarters of a million pairs sold out almost at once. 'Is Nylon's Future the Past of Silk?' speculated the *Department Store Economist* – not an unreasonable assumption when, by 1941, nylon hosiery was selling from $1.25 to $2.50 a pair while silk had dropped from $1.35 to $1.00.[50]

As a marketing strategy, nylons were often displayed in the context of scientific apparatus and the raw materials of the fabric – the magic of it all must not be forgotten – and hordes of shoppers immediately emptied the shelves within hours of every delivery. The important underlying psychological shift that this signified was clear as early as 1942: when American women accepted nylon's embrace, they also gave up 'the unpleasant connotations that have so long clung to the idea of a chemical substitute'.[51] Nylon had rapidly invaded the lucrative feminine markets of stockings, corsets and lingerie and by November 1939 it was making inroads on the equally profitable man's world of socks and ties.

Unlike silk, nylon was stable in supply, price and quality. It created a climate where design was encouraged: new materials with price stability led to more new products and more ways of selling them. Before nylon there were few shops specializing in hosiery but within a few months roadside kiosks or nylon bars were springing up along the highways.

THE NEW STOCKING WITH THE BLUE STRIPE
Miracle
NYLONS
NOW!!
Extra Fine
DRESSY
SHEERS
NYLON
$1.00
FIRST QUALITY
FULL FASHIONED
ALL DU PONT NYLON
AND FULLY GUARANTEED
Easy to buy – Lovely to wear
4 QUARTERS
CLOSE
CONNECTION
RUSH
23

Airport vending machines dispensed nylon stockings and they were sold by cigarette girls in smart city nightclubs.[52] In 1940, the *Du Pont Magazine* told its readers that it had already invested $28 million in nylon manufacture and that 90 per cent of the nylon produced was going into women's fully-fashioned hosiery. However, to demonstrate nylon's unprecedented strength, abrasion resistance and elasticity, it was made into football pants for a game at the Yankee Stadium just one month before the Japanese attack on Pearl Harbor in December 1941 – the shocking provocation that drew the United States into the Second World War.[53]

Nylon at War

When America entered the war, the War Production Board immediately announced plans to take over all nylon yarns for military purposes and nylon formally went to war on 11 February 1942 when Du Pont turned all its production over to the war effort. A nylon flag was made for the White House, symbolizing America's ability to do without Japan and its silk. Until V-J Day the only nylons to be found were those that were produced from pre-war stock. With supplies of silk cut off, heavier nylon yarns were needed for making parachutes, the canopies of which alone took about 65 metres (210 feet) of fabric. Nylon was also made into ropes, tyre cord, tents, uniforms, mosquito netting, tarpaulins, hammocks, webbing, bomber noses and shoelaces.

The military found more and more uses for nylon fibre: it was resilient, strong, resistant to mildew and had no attraction for the usual pests like clothes moths. They could not get enough of it and the sacrifice of her nylons became a woman's great patriotic gesture. Newspapers reported stories of girls 'Sending Their Nylons off to War', carried photographs of leggy women taking their stockings off 'for Uncle Sam' and described how it would take 4,000 pairs to make two bomber tyres. In 1943, patriotic women turned in 7,443,160 pairs of silk or nylon hosiery for the war effort, and Betty Grable, who owned the most famous legs in wartime America, took off her nylon stockings at a War Bond auction and offered them for sale. The successful bidder paid an astonishing $40,000 for these icons of American glamour and progress.[54] Even the few people who had not heard of nylon before suddenly realized that this new chemical product was playing a vital part in their lives. Because the war had cut off or curtailed imports of wool, silk and linen, there would have been clothing shortages were it not for the fact that 'The American chemical industry could be depended upon to do its part in keeping our soldiers and sailors well clothed for service, and the civilian population, too.'[55]

When Nylons Bloom Again

The 'nylons' issue was an international preoccupation during the war, heightened by the fact that they had only just become available before they disappeared. Nylons became currency in black-market circles, selling for $10 or $12 a pair, and women resorted to all kinds of deceptions to create the illusion of wearing nylons. Legs were dyed and wobbly lines emulating the stocking seam were drawn down the back of thighs and calves. Women longed for the day when they could buy nylons again. A song, 'When the Nylons Bloom Again', with lyrics by George Marion Jr and music by Fats Waller, was performed in the show *Early to Bed*, with the chorus 'I'll be happy when the nylons bloom again, cotton is monotonous to men'.

Within days of Japan's surrender, on 22 August 1945, Du Pont announced that it was re-directing its nylon production and that it would be available in stockings and civilian dress by September. The symbolic power of a pair of nylons was such that this statement was like a starting pistol at a racetrack. 'Nylons by Christmas!' became the country's motto and Du Pont forecast that there would be enough nylon manufactured to knit 360 million pairs of nylon stockings per year – 'the equivalent of eleven pairs for every American woman'.[56] As it turned out, this was an over-optimistic estimate as technical delays held up volume production. When the limited number of post-war nylons did finally appear, in 1946, the reaction was riotous. In Chicago, a headline trumpeted 'Peace, It's Here! Nylons on Sale'.

Opposite:
Left: Nylon was soon as much a part of the American scene as baseball and football. In 1941, the first nylon football pants were made.
Right, top: The perfect military material, nylon, went to war. All nylon, including reclaimed hosiery, was dedicated to the making of parachute canopies and other vital wartime items.
Right, bottom: Betty Grable at a War Bond rally where her stockings were auctioned for an amazing $40,000.
Background: The first nylon flag: in 1942, a Du Pont nylon flag was hoisted over the White House, a symbol of America's new-found independence from the Japanese silk industry.

news about
NYLON
SNOW'S FUN for Junior—and Mother's happy, too, when his snowsuit's nylon! Warm, sturdy, long-wearing—strong when wet—light and comfortable—nylon fibers are just as hard to wear out as Junior.
BLOCKING, TACKLING, in the mud, on frozen ground—football players give their uniforms plenty of pounding. That's why light, tough, easy-to-clean nylon has its own cheering section. And how nylon wears—lasts through several football seasons!
Du Pont makes the nylon fibers used in the products shown. The manufacturers of these products use nylon because nylon products can have these outstanding properties:
√ STRENGTH
√ LIGHTNESS
√ ELASTICITY
√ TOUGHNESS
√ LONG WEAR
√ EASY WASHING
√ FAST DRYING
√ FLAME RESISTANCE
√ RESISTANCE TO MOTHS AND PERSPIRATION
√ CAN BE "SET" TO HOLD SHAPE
DU PONT
REG. U.S. PAT. OFF.
BETTER THINGS FOR BETTER LIVING ...THROUGH CHEMISTRY
TO MANUFACTURERS: Are there textile fibers in your product? Then you'll want to read "Nylon Textile Fibers in Industry." Write for this important new book!
NEW! FREE book for women, teachers, students—"About Du Pont Nylon." Write to Nylon Division, E. I. du Pont de Nemours & Co. (Inc.), Wilmington 98, Del.
YOU TRAVEL LIGHT when you travel with nylon. Light, strong, nylon fabrics for luggage make bags easy to carry, cleanable, tough to scuff or rip. Good travel companions, hard knocks won't hurt their service record.
RIDE 'EM, COWBOY! Roping and tying put terrific strain on a lariat—but nylon can take it. Nylon "gives," takes a shock without snapping. Nylon is tough, elastic, not weakened by mildew or soil rot. Try Nylon.
IT'S DREAMSTUFF—sheer-but-strong nylon tulle. Hard to tear, but so light and lovely. Your gown keeps fresh and crisp all evening, because nylon can be "set" to hold shape. When a dress looks lovely longer, chances are it's nylon!
FOR NYLON... FOR RAYON... FOR FIBERS TO COME... LOOK TO DU PONT

When nylons reappeared after the war, customer lines made merchandizing history. During these 'nylon riots' many buyers fainted in the crush.

The *New York Times* reported '30,000 Women Join in Rush for Nylons', and in Pittsburgh the story was summed up with these words: 'Nylon Mob, 40,000 Strong, Shrieks and Sways for Mile'. For months newspapers carried headlines and pictures of the long lines of women waiting to buy nylons.

The 'nylon riots' began in September 1945 and continued until the middle of the following year. The day-long queues and scenes of near hysterical, brawling women were gleefully reported in the press: 'Women Risk Life and Limb in Bitter Battle for Nylons', or 'Nylon Sale and No Casualties'. On the occasion when 40,000 people queued to compete for 13,000 pairs of stockings, the Pittsburgh newspaper reported 'A good old fashioned hair-pulling, face-scratching fight broke out in the line'. According to the *American*, 'the mournful news about nylon shortages should cause women's pages from coast to coast to be bordered in black'. But, as this journalist observed, no other commodity had ever received so much free advertising. The scarcity of nylon stockings had turned them into symbols of luxury, and the passion with which they were pursued could almost be likened to the Beatlemania of the next generation. Some people began to suspect that Du Pont, who had the monopoly on manufacturing rights, were deliberately holding back production for the publicity, although this allegation was unfounded.

The Fabricated Future Arrives

By the end of 1946, newspapers were reporting that nylon queues were getting shorter and even disappearing in some cities. The 'miracle' called nylon, Du Pont acknowledged, was becoming commonplace, just as brand new 'miracles' such as 'Orlon' acrylic, 'Dacron' polyester and 'Lycra' spandex fibre were being mixed in the Du Pont test tubes.[57] It was the consumer, and specifically female consumers, who accelerated the acceptance and production of chemical textiles, but it is interesting to speculate to what degree the wartime scarcity of nylon lent it mystique and value. When nylons finally appeared in Britain after the war there were signs in shop windows saying 'only available to foreign visitors' and passports had to be shown before a pair could be bought. They were costly enough for Sketchleys, the dry cleaners, to employ girls specifically for the purpose of knitting up the ladders in nylons; they were placed conspicuously in the windows to work.

By 1946 nylon was 'as well known to the American public as baseball' and it seemed that a more egalitarian future would be made possible because substitutes for the old raw materials had been found. 'Nylon is returning to serve you' Du Pont announced in 1946, and to stimulate the public imagination and illustrate nylon's versatility in tangible form, a Nylon Showroom was opened in Wilmington. Here, wedding scenes were exhibited, including the bride's boudoir, with nylon transparent curtains at the window and nylon satin bedding, and the Nylon Bride herself, dressed in nylon and with an all-nylon veil and flowers, with a 'colorful array of gloves, girdles, slips, scarves, lingerie, blouses and hosiery'.[58] The Nylon Merchandise Laboratory presented these fabrics and garments to show the limitless opportunities polyamides would bring. Before long, nylon spread not just to the rest of the wardrobe but into home furnishings – into carpets, curtains, upholstery and bedding (soon to be followed by the other major synthetics polyester and acrylics). Three hundred years of fibre experimentation and chemical speculation were about to pay off and the fibre producers and synthetic-textile manufacturers would enjoy unparalleled profits for the next three decades.

This textile-fibre revolution came not from the fabric but from the chemical industry. Perhaps too attached to traditional materials and locked into the machinery and dyestuffs that functioned only with natural fibres, the textile manufacturers had no real impetus to re-invent their stagnating industry. It was the massive petrochemical conglomerates, particularly Du Pont, that dragged the fabric manufacturers into the twentieth century and the irresistible mechanism by which they achieved this was by linking synthetics to fashion. By targeting women and using a product that spoke volumes about material fragility and sexual vulnerability – stockings – the implied threat of technology was utterly neutralized. The chemists were modern day Merlins, working in the service of and for the ultimate good of an innocent pleasure: fashion. It was fashion, therefore, that created a compelling demand for synthetic fabrics and the textile manufacturers responded to this powerful force; they stepped into the arena and, during the 1950s and early 1960s, did everything they could to promote synthetic textiles within a fashion context. It was in 1939 that America's fashion consumers first met the synthetic chemist. In the decades that followed, first branded or named fabrics and ultimately designer labels would be encountered on synthetic clothes.

The first completely synthetic fibre, nylon, was a material without a past or an identity and – if any lessons can be learnt from history – the story of polyamide (and its chemical acquaintances polyester and acrylic) graphically illustrates the powerful impact of a good marketing campaign. Selling synthetics is selling the invisible and depends upon the power of seduction, convenience, performance, function, economy and, most of all, luxury. 'Wonder' and 'magic' were frequent prefixes to nylon fibres, but it was 'miracle' that was used most often.

The synthetic textile fibres which Du Pont so prolifically fostered gave clothes designers a new world of cloths – a timely entry to the colours, shapes and textures of modernity. It seems impossible to imagine synthetic textiles being invented anywhere other than in America. Although the traditions of cloth-making were thousands of years old and the industrial revolution in Europe mechanized the textile industry beyond recognition, there was essentially something missing which gave European fabric innovation its stop–start character. Technology, in the textile context, was historically a dangerous and destabilizing beast and there was a certain wariness about breaking with tradition. The way in which

Du Pont's vision of married bliss in nylon. The Nylon Bride in her boudoir, 1944.

Courtaulds adopted rayon purely because of its similarities to silk, the fibre it already held so dear, indicates a lack of appetite for the new, although, ironically, it was rayon's success that shifted the manufacturing emphasis at Courtaulds away from textile manufacture and towards chemical-fibre production. In the United States, however, there was no conflict between the idea of newness and that of traditional culture. America's frontier mentality made it the natural home of reason-defying technical feats.

To Henry Ford, technology was an American religion and his influence was enormous in the inter-war years. American scientists fought off the effects of price fluctuations and of geological monopolies in organic raw materials simply by re-inventing them in the laboratory. They had developed the most efficient way of making standardized goods available to the mass market. Consumers already drove around in assembly-line cars and lived in prefabricated homes with injection-moulded products; now they would enjoy the pleasures of fashionable synthetic clothes – more things for more and more people.

3

1950–60

Better Living through Chemistry

'The consumer isn't a moron; she is your wife.'

David Ogilvy,
Confessions of an Advertising Man

'Better Things, for Better Living – Through Chemistry.' Du Pont's famous pre-war advertising tag became a ready-made slogan for the consumer boom of the 1950s: the decade when prosperous America taught the rest of the world how to find happiness – through *things*. Plastics came into their own with the acceleration in demand for more and more products and materials to furnish the homes and fill the wardrobes of newlyweds, and synthetics provided the material answer for the easy-care fashionable clothes demanded by housewives and the recently discovered 'teenagers' alike. Immediately after the war, it was the 'housewife' who became a central marketing target for the mighty petrochemical giant Du Pont – 'What really makes Du Pont, the company, live and breathe is – women!' declared a 1947 report in *Inside USA*, 'Its history is a development from dynamite to nylon. By a strange paradox, the women, not the men of the world, are the ultimate determinant of Du Pont policy. Much more than on dynamite, the company rests on housewives.'[1]

As new markets were identified, prototype garments began to emerge, marking the point at which the synthetic fashion revolution was really launched. After centuries of chemical experimentation, nylon arrived in the form of stockings in 1939, and the widespread excitement about this new fibre was only increased by its disappearance during the war. With the arrival of even newer fibres and experimental garments came a flood of fibre marketing, pointedly targeted at young and married woman. Newly branded fibres and fabrics were sold with an unprecedented intensity by the competing chemical corporations and this chapter will look at the methods by which consumers were won over to nylon, polyester and acrylic and at how the synthetic fibre manufacturers expanded their markets out of hosiery and into mass fashion.

Opposite: As Du Pont chemists weave a revolution in lightweight men's suitings, a happy couple step into their easy-care future in 1950.

The elated American housewife in her new consumer paradise of 1952. The dream comes true.

The Synthetic Shopping Utopia

As the 1950s unfolded, it became clear that synthetics and plastics were to be the materials that would literally shape and feed the growing design boom and which would introduce 'a world of plenty' to both the wardrobe and the home. America symbolized progress through abundance and happiness bought with new products – the 'American Dream' which was adopted so enthusiastically in post-war Britain. The Britons described by J. K. Galbraith in *The Affluent Society* [2] were only too eager to imitate the 'populuxe' lifestyle of their transatlantic neighbours and the material basis of everyday life was about to be transformed – thanks to the wealth of man-made materials. Synthetics (and plastics) were the agents behind the obsolescence that lies at the heart of the modern fashion industry – mass produced garments, made from artificial substances, that can be discarded as casually and as easily as they are purchased.

A unique set of circumstances consolidated the mass production of fashionable clothing and boosted the widespread consumption of man-made materials. At one and the same time there was a shortage of all traditional, natural manufacturing materials, a seemingly unquenchable thirst for more and more things to consume, and an increase in youthful affluence which particularly intensified the demand for new clothing. This was exactly the climate within which the newly invented synthetic fibres could flourish and in the 1950s synthetic-fibre producers regarded themselves as responding heroically to the dramatic increase in textile demand. The privations of war and continued rationing had fired a widespread appetite for new clothes, added to which natural fibres were prone to severe price fluctuations. Therefore, it is not surprising that science was perceived as the gateway to the shopping utopia. If nature could not provide it, then the chemist would.

Nylon was a fibre unlike any other and with qualities beyond the 'big dream of the old inventor', that of simply simulating nature's fibres. Instead, Du Pont measured its progress by the 'sharp distinction between new fibres and imitation' and unless a new fibre possessed the 'easy living' qualities of nylon it wasn't 'worth more than an honorable mention in a routine laboratory report'.[3] As soon as the war ended, there was increasing pressure on Du Pont to licence the manufacture of nylon. Du Pont had already granted ICI a licence to produce nylon in Britain and the Commonwealth and, because they had never patented the name 'nylon', the right to use it as the generic name of the fibre went with their licences. The first of the many licences granted in the United States went to the Chemstrand Corporation in 1951.

Polyester Reigns

During the fifties, a stream of new synthetic fibres almost overwhelmed the textile market. Not just nylon but acrylic and, most importantly, polyester, for, despite nylon's fame, it was polyester that became the most important synthetic fibre in terms of volume and garment production.[4] Although Carothers and Hill at the Du Pont laboratories had made the first inroads towards the discovery of polyethylene terephthalate (polyester or PET) in 1931, it was fully developed by two British chemists, John Rex Whinfield and J. T. Dickson. The Carothers–Hill version of the fibre was, they found, both too expensive to manufacture commercially and had a melting point that was too low to make it suitable as a textile fibre. It was at the laboratories of the Calico Printers Association in Accrington, Lancashire, that Whinfield and Dickson finally made the first test-tube filament of polyester in 1941.[5] In 1946, the American manufacturing rights were sold to Du Pont and the world rights to ICI. These chemical giants worked out their own manufacturing processes. The first public announcement of ICI's brand 'Terylene' was placed in the *Manchester Guardian* on 5 October 1946. This, along with Du Pont's polyester 'Dacron', would soon command vast world markets. British clothing restrictions were lifted in 1952, just in time for the introduction of Terylene fashions, and by the mid-1950s, large-scale manufacture of polyester fibres was under way in Wilton, Teesside and, from 1954, at Du Pont's new plant in Kinston, North Carolina.

Nylon, too, was put into accelerated peacetime production. In 1940, Courtaulds and ICI had formed a jointly-owned company, British Nylon Spinners (BNS), although during the war all British nylon went for military use. BNS became the cause of a sometimes troubled partnership between Courtaulds and ICI. In the post-war period, ICI came to dominate Britain's synthetics industry. Like Du Pont, it was a multi-divisional corporation manufacturing a wide range of chemically-based products, from pharmaceuticals, to plastics, paints and fertilizers; its Fibres Division was just one of many. The great advantage of what amounts to a vast chemical republic is that by-products can be re-formed into yet another saleable product. Already a leading supplier of chemicals to the textile industry (dyestuffs and pigments, caustic soda for the manufacture of rayon, bleach, zip fasteners and so on), the introduction of synthetic fibres greatly increased ICI's focus on textiles. Conscious of their enormous chemical opportunity, ICI set about building up their Synthetic Fibres Division until it became the British equivalent of Du Pont's. By uniting Courtaulds and ICI, nylon fibre forced the coming together of textiles and chemicals.

Petrochemical Power

It was the task of the burgeoning fibre manufacturers to invent a demand for their previously unknown brands and economic and social factors made the time ripe. It was not until after the war that the ready-to-wear fashion industry reached sufficient manufacturing and retailing maturity to fully exploit both the new synthetic materials and the popular

Above: Men's socks represented massive market fortunes in fibre sales. Nylon, the drip-dry option, invaded male hosiery in the 1950s. Below: A 1944 advertisement shows the scarves, artificial leather and bombs that are all created from 'Toluene', a component part of the high explosive TNT made at Shell's University of Petroleum.

demand for mass fashion. Higher incomes, consumer credit and the escalating production of manufactured goods ensured that synthetics were gratefully welcomed by manufacturers and consumers alike during the expansive years of the 1950s. But this proliferation of synthetic goods was only made possible by the growing range of raw materials that fed the industry, and the most significant factor in this was the switch from coal and gas to oil. In the nineteenth and early twentieth centuries, the world's plastics and organic synthetics industries were based on four main raw materials: coal, limestone, cellulose and molasses. After the war came the critical switch to oil. Synthetic fibres were to become products of the petrochemical industries and, as one of the world's most oil-rich nations, the United States was ideally placed to blaze a trail in both the making and marketing of its chemically created materials.

Alec Guinness played an idealistic scientist in *The Man in the White Suit* in 1951. In fabric-starved Britain, he invents a polyester-like 'indestructible' suit and lives to regret it.

Not only benefiting from their abundance of petroleum, America's synthetic manufacturers had also witnessed the obliteration of their potential rival, the Japanese silk industry. *American Fabrics* described in 1950 how the silk producers had underestimated the threat of man-made fibres:

'Of all the natural fibre fields which were invaded by the development of synthetics, none showed a greater apathy or a lesser willingness to protect its future than the silk industry. The producers of cotton and wool were quick to recognize the potential competition of test-tube-born fibers, and equally quick to develop programs of education as well as merchandising strength to retain and even expand the end uses of their products in competition with synthetics. The silk industry, however...from the cultivator of mulberry trees in the Far East to the finisher of fabrics in Paterson, played hare to rayon's tortoise. Imbued with the false sense of security which has always been aristocracy's bête noire, the producers of silk at first refused even to concede the existence and worth of synthetic fibers. By the time rayon had been refined in product and in promotion to the point where it was first-line competition, the beginning of World War II deprived American silk companies of their source of raw material and thereby of their weapon of combat. For, during ten long years, without actual silk in hand it became a matter of pitting words against fabrics.'[6]

During the war years the spinners and weavers of silk had turned almost exclusively to the use of rayon and many of the specialist skills and techniques of working with silk fibres were lost to the industry over the course of a decade. Equally important, as *American Fabrics* acknowledged, was 'the loss of a complete generation of the consuming public'. Men and women aged between 21 and 30, some 20-odd million of rich consumer potential, 'hardly knew silk, except as a word – it wore little silk, felt little silk' and 'its standards of excellence were not based on silk'. Optimistically, *American Fabrics* believed that the new niche for silk's rehabilitation would be in representing it as the 'class product for the masses'[7] – this position, however, was to be claimed for nylon.

Not just technology and fibre innovation but product differentiation and brand patenting meant everything in the hugely competitive textile industry. For the first time, the multinational petrochemical companies involved in the research and development of synthetic materials, had to confront the ephemeral world of fashion and the problem was how to substantiate the inherently invisible qualities of synthetic fibres. 'You can't sell 'em what they can't see' was how *American Fabrics* put it in a 1951 article entitled 'Promoting A Hidden Product'. Synthetic-fibre manufacturers had a daunting task: they had to win over spinners, weavers, knitters, garment manufacturers, designers, retailers and, most important of all, consumers. Manufacturers set about

Polyester was a serious threat to Britain's famous woollen industry. A Hong Kong tailor makes a traditional 'Savile Row' woollen suit, 1963.

Synthetic suits were the marvel of the early 1950s. Du Pont's frequent promotional stunts demonstrated their wash-and-wearability, including salesmen leaping cheerfully into swimming pools.

Du Pont presents the wonders of machine-washable clothes. A ladies' suit with a fashionable permanently-pleated skirt is put to the test.

the task of winning the hearts of housewives, bachelors and travellers – based on the labour-saving and convenience qualities of new materials. Synthetics were presented as being modern and highly fashionable, but at first convenience was the advertising priority. Polyester was to give the world 'the man in the wash-and-wear suit'.

Automatic Suits: Man's Modern Marvel

In 1951, Du Pont launched their polyester, 'Dacron' (formerly known as 'Fiber V'). This was the magic material that promised to make ironing obsolete. That same year, a British film, *The Man in the White Suit* (starring Alec Guinness in the role of the definitive chemist-hero), was released. This famous fable tested the post-war dream of a clothing utopia made possible through chemistry. It tells of the ultimate wonder fibre – an indestructible fabric that stayed clean forever, never knew an iron and which, its idealistic hero believed, would bring an end to the exploitation of textile and clothing workers and to the endless laundering drudgery of housewives. But all these good intentions soon fell into ruins when it was shown that the actual end product of his new fibre was mass unemployment. Finally, the disillusioned scientist became a laughing stock when his prototype suit fell to pieces – proving that modern textiles were as unreliable as a blind faith in science. The dangers of too much progress were thus averted, the status quo restored, and the daring chemist was returned to the obscurity of his laboratory.

The Man in the White Suit was, in reality, symptomatic of many of the attendant worries of the sudden chemical-fibre revolution. Nothing would ever be the same again: textile making, clothing manufacture, retailing, dry cleaning and, most confusing of all, the social status and value attached to fabrics and fashions. Would an everlasting suit be the dream or the nightmare of modern man? This was the cultural backdrop against which polyester was set. Polyester was to be the single most revolutionary fibre in the shaping of contemporary menswear and the textile that would rapidly descend from modern miracle to joke status within just one generation.

Du Pont literally made a big splash with the launch of Dacron. Sold on its 'wash-and-wear, eases care' merits, Du Pont delighted in publicly demonstrating these qualities by having Dacron-suited salesmen jump into swimming pools or emerge dripping from showers or baths. In 1951, the *Du Pont Magazine* published a report on its testing of the first experimental all-Dacron suit. 'A suit of 100 per cent "Dacron" was worn for 67 days without pressing. During that period the owner literally went overboard in the cause of science. He wore the suit twice in a swimming pool, washed it once, and found it still

PORT OF LONDON AUTHORITY
LAMBE
Wash and Wear "DECTON," Photographed for Arrow in London
Wherever you go you look better in an Arrow Decton shirt...
The fabulous Arrow DECTON shirt . . . of "Sanforized" wash and wear fabric that's smooth, luxurious . . . and so comfortable 52 weeks out of 52. And what durability! DECTON outlasts any all-cotton shirt, because it combines 65% DuPont polyester fiber and 35% long-staple cotton. In white, colors, stripes; favorite collar styles. From 5.00. All-silk tie, 2.50.
ARROW
TRADE MARK

presentable enough to be worn to the office.'[8] Men's wash-and-wear suits were launched in 1952 in blends of acrylic, polyester and nylon.

In its infancy, Dacron captured public imagination because of its two distinctive characteristics which textile scientists called 'moisture insensitivity' and 'rapid recovery from stress', both increasingly important qualities in changing post-war lifestyles. Synthetics could bring an end to the ancient upkeep problems of hot-weather wardrobes and modern living was bringing in a 'trend to comfort'. With civilization providing well-heated offices, cars and homes, the lightweight suit could conveniently be worn all year round. The old heavyweight winter suits of 14- to 16-ounce (400–450 g) fabrics were 'going the way of the top hat' and in their place were synthetic fabrics weighing as low as 11 or even 8 ounces a yard (310 and 225 g per metre).[9] Soon suitings were being made in 6-ounce weights (170 g), so that a 1959 suit weighed less than half that of one made in 1939. Sales of synthetic suits were climbing steeply. In 1958 it was estimated that 2.3 million men's and boys' synthetic-fabric suits and more than a million slacks were sold in the United States; sales of synthetic garments in total – men's, women's, boys' and girls' – were valued at more than $1,500 million.[10]

'Travel Light.... And Love It' advised *Gentry* magazine in 1956, when air travel and tourism were beginning to become more commonplace. The strict baggage allowances imposed by airlines was another very good reason for turning to synthetics, and, as *Gentry* magazine 'humbly suggested', 'it's possible for a woman to be well dressed on a trip without taking everything in her closet...here again, the man-made fibers provide maximum use from relatively few items with a minimum of care...."Living out of a suitcase" doesn't have the bad connotation it used to, thanks to the kind of carefree clothing you can put in a suitcase nowadays.'[11]

Polyester was also presented as a hard-working fibre and, in 1955, *Du Pont Magazine* published 'Dacron on Duty', an article illustrating the variety of uniforms that were made from Dacron or Dacron blends for state troopers, firemen, policemen, U.S. Navy officers, airline stewardesses, gas-station dealers, railway guards, water-meter readers and repairmen – all of which 'stayed neat' because of the new polyester.[12]

Then there were drip-dry shirts, and a prime target for these was the travelling salesman. Market researchers at Du Pont had pinpointed a growing social and career problem for him: 'Moving from town to town, arriving late at night in distant hotels, he was completely out of step with U.S. laundries. Business instincts, buttressed by surveys, suggested he'd be interested in easy-care shirts. He was. The shirt-washing ritual in hotel rooms became almost as common as dinner hour.'[13] Brooks Brothers were the first to experiment with a 60 per cent polyester/ 40 per cent cotton shirt (still today's familiar combination). This was, according to *American Fabrics and Fashions,* 'a marriage made in heaven' because it brought together the performance of the man-made and the 'natural hand' (the feel of natural fibres).

'Wrinkle Studies' were among the many intensive tests carried out by Du Pont's textile specialists before the concept of 'automatic wash-and-wear' was finally introduced in 1956. Newspaper headlines urged ladies to 'lay down your irons' because, in the era of synthetic garments, everything could now be washed and dried in home machines and would emerge virtually wrinkle free. Du Pont organized many demonstrations of tailored jackets and pleated skirts being put through the home washer. 'This spring the whole family can wear clothes made to be washed and dried – wrinkle-free – entirely by machine.'[14] Between 1950 and 1956, the sale of washing machines in the United States more than tripled.

THREE WASH-AND-WEAR SUITS: Haspel seersucker ("Orlon" and cotton), Witty tropical (all "Dacron"), and Schoeneman cord ("Dacron" and nylon). Du Pont fibers help the suits retain a press despite laundering.

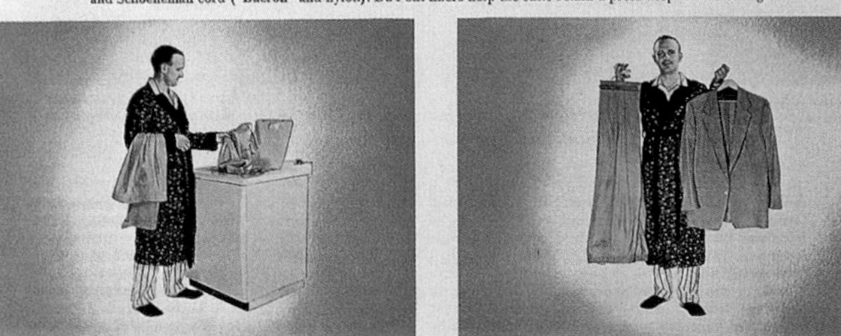

INTO THE SUDS goes the Schoeneman suit. It's put through the washing and rinsing cycles, but *not* the drying step. Instead, it's suspended from hangers and allowed to drip dry. All three suits should be washed this way.

DRIED OVERNIGHT, the suit can be worn again the next day without pressing, as shown here. Wash-and-wear suits also resist wrinkling, are ideal for travel wear. They'll be on sale this year, although in limited numbers.

Above: The miracle of the wash-and-wear suit in 'Orlon', 'Dacron' and nylon. Throw it in the washing machine then let it drip dry overnight, a boon to the travelling salesman.
Opposite, top: Wash-and-wear 'Decton', a drip-dry shirt in the classic blend of 65 per cent polyester and 35 per cent cotton, photographed for Arrow in stylish London in 1960.
Opposite, bottom: 'Fortrel', a brand of polyester fibre, was illustrated in a glamorous fashion context in the *ICI Magazine*, 1963.

'The washable suit is the suit of the future' claimed the well-known New York suit manufacturer, Spencer Witty, whose favourite means of proving this point was to pour Indian ink, iodine and ketchup over a pair of trousers – before his startled customers – and then to 'toss them in a washing machine and hang them up to dry'. After just one hour, Witty would lead his visitors back to see the trousers, 'nearly dry, spotlessly clean and still in press'.[15] The 'Witty Tropical' was an all-Dacron product but other suits with shape-and-press retaining properties were manufactured using nylon, acrylic and polyester blends. (It was important for the lapel padding, lining, pockets, seam tape and sewing threads to be synthetic so that 100 per cent washability could be ensured.) In the humid states of America, the neat wrinkle-resistance of synthetics was highly valued, particularly among businessmen. Consumer reaction was overwhelmingly favourable, despite the fact that these suits were introduced at a relatively high price, anything from $40 for a synthetic seersucker to $100 for a Witty Tropical, but – as Spencer Witty pointed out – prices would come down and production would be stepped up. He was right; every cotton producer in the textile industry was soon affected by the arrival of polyester and many moved the bulk of their production over to the new synthetics. Other fibre producers, too, 'saw the rosy future' that polyester was opening up and hurried into production; Eastman introduced Kodel in 1958 and Celanese launched Fortrel in 1960, in conjunction with Britain's ICI.[16]

'Clothing today is big business', announced *Better Living* in 1959, employing 1.5 million men and women and with $9.4 billion-worth of apparel being sold in the previous year, much of which incorporated 'the innovations of the past half dozen years'.

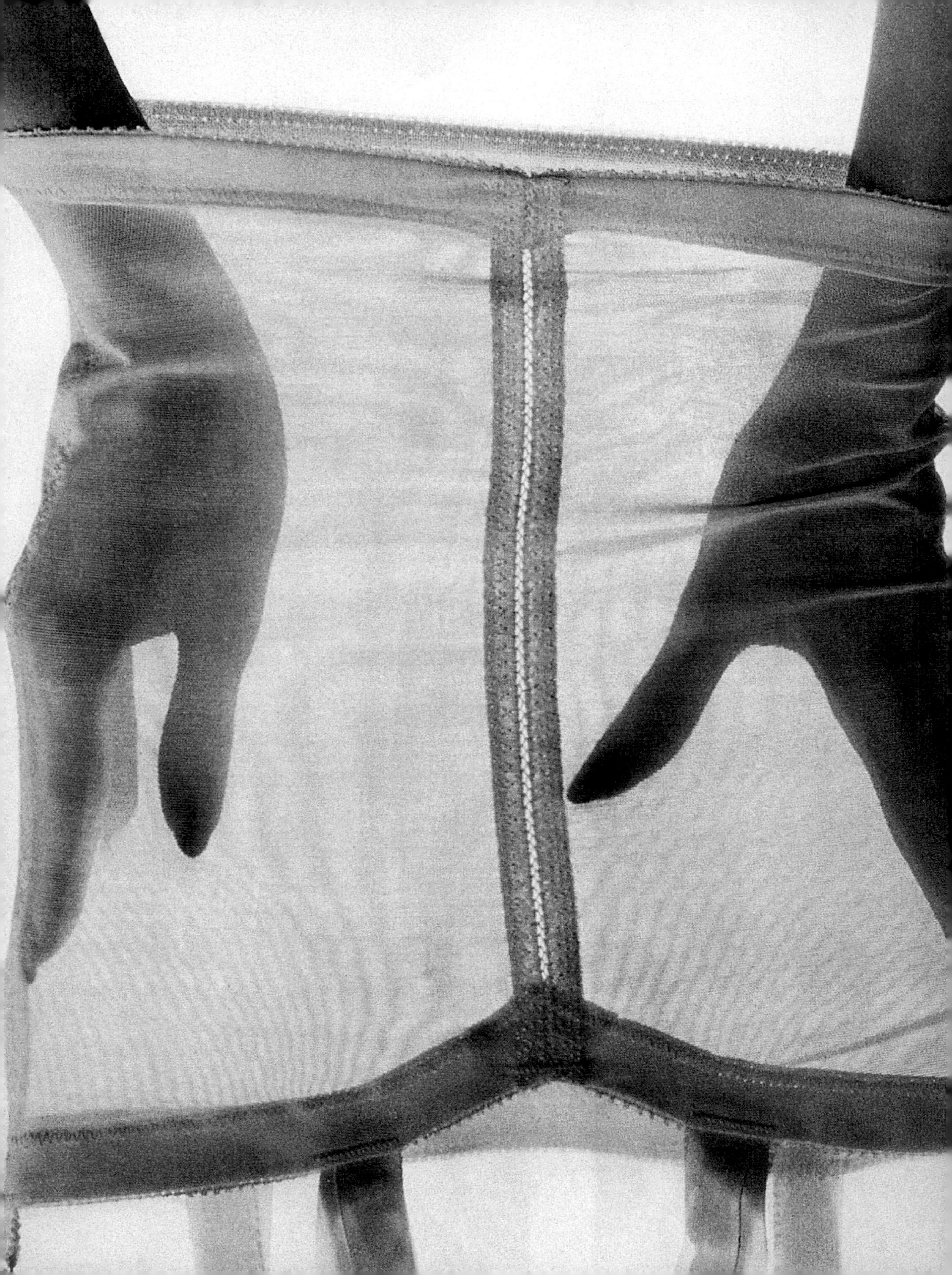

There was every indication that the synthetics revolution was unstoppable: '"We are living in one of the most exciting textile ages man has ever known", summarizes a textile mill executive. "Each year brings technical gains that dwarf those of many previous decades or generations. It's plain that the big research and development work of Du Pont and others in the past 20 years is fostering a sweeping transformation in U.S. clothing habits...to be seen all around us."'[17]

The man-made fibres created in industrial laboratories were the 'prime factor in the twentieth-century wardrobe revolution'. Synthetics had brought in their own style changes and all the drabness and formality of the previous generation was being replaced by 'color, gaiety and informality'. Cumbersome and uncomfortable garments were now outmoded and variety, not monotony, characterized wardrobes. Uniquely, the qualities of man-made fibres had 'pioneered a new concept in clothing, that of engineering all fibers and fabrics to meet consumers' specific and changing needs.'[18] One of the exciting new products then under way was 'Fiber K', which later became the world famous brand 'Lycra' – a synthetic elastic with more strength but much less weight than natural rubber yarn. Announced in 1959, Du Pont carried out extensive garment tests on its new stretch fibre and in one early demonstration a lady's girdle made from the fibre was dramatically unfurled from a cigarette packet.

Above: Dresses featured in *The Big Payoff*, a Du Pont-sponsored television show broadcast in the United States. A teenager shows a sun-ray pleated dress in 1957.
Opposite: 'Lycra', Du Pont's 'light and mighty' synthetic spandex fibre, introduced in 1959, made sheer, figure-controlling girdles possible. They were a third of the weight of conventional foundation garments.

Motion-picture Marketing

Beyond research and development came promotion and in the decade following the end of the war, more and more resources went into market research and advertising campaigns to win over the new generation to the blessings of plastics and polymer fabrics. The Du Pont advertising machine made use of every new media to reach its clients – even if the finished products were three or four times removed from the Du Pont raw materials. Radio, television and film were vehicles of fibre news and product information. *The Cavalcade of America* was Du Pont's sponsored television show and its own Motion Picture Library has many examples of films made to promote its fashion fibres and including technical processes never before shown on the screen, 'taking the audience to the elbow of the chemist as he works in his laboratory'. Hollywood actors and actresses were cast in *A New World Through Chemistry* which takes its viewers on 'a magic carpet ride' through manufacturing plants and into the 'the boudoir that chemistry made'. 'The setting is enlivened by a Hollywood actress dressed from head to toe in nylon. The rugs, draperies and decorations are all made of materials born in the laboratories. Even the furniture is made of transparent "Lucite" methyl methacrylate resin. This picture also shows several stages in the production of nylon [and] how neoprene is derived basically from coal, limestone and salt.'[19]

Included in the Motion Picture Bureau's output was *The Wonder World of Chemistry* which reviewed the progress of products from laboratory to daily use, showing how cellulose, the framework of plant life, is transformed into colourful plastics and rayon fabrics. *This is Nylon* featured Hollywood players (a showgirl and four male singers) glamorizing the new synthetic, while *Starring Nylon Stockings* was made in 1952 by the Textiles Fibers Department to show department-store buyers and customers how to select nylon hosiery for a smooth fit. But the popular fascination with all things chemical is perhaps best expressed in United Artists' early 1950s film *Eternally Yours*, starring Loretta Young and David Niven, where David Niven, an illusionist, 'creates' Loretta Young from mysterious chemicals.

Educating the consumer about the coming synthetic wonders was an underlying theme of countless magazine articles and, in many cases, editorials. In 1947, *House Beautiful*, in collaboration with the American Society of the Plastics Industry, published 50 pages illustrating that plastics were the 'Way to a Better, More Carefree Life'. 'Although it was "all very well for the chemists to talk of their molecules and polymers, and for the engineers to carry on about extrusions and injections", there was "only one good reason why you, personally, should be interested in plastics." That reason was "damp-cloth cleaning" – a concept "about which you'll read constantly in the pages that follow".'[20] *American Fabrics and Fashions* hailed the advent of man-made fibres as 'a liberation from domestic slavery'. 'How strange', wrote its editor in 1985, 'that today not a single eyebrow is raised in astonishment or praise of the man-made fibre that brought us the miracles of wash-and-wear, permanent press and pleating, wrinkle and soil resistance. No tombstone marks the end of an era in which women were tied to a washboard and iron as an almost daily duty. Yet the tremendous change in our lives brought by easy-care clothes is visible everywhere.'[21] Plastics and easy-care fabrics would rescue hapless wives 'from the ravages of dogs, kids and husbands', promised *Better Homes*.[22] The latent message behind this was that domestic work was solely a feminine concern. It was, however, their role as the purchasers of domestic products, and particularly clothing, that made women so crucial to the multi-million dollar synthetic-textile industry. The new range of synthetic fabrics were seen as 'miracle materials', hygienic, modern and labour saving and somehow symbolic of the buoyant scientific future seen at the New York World's Fair and the Festival of Britain in 1951.

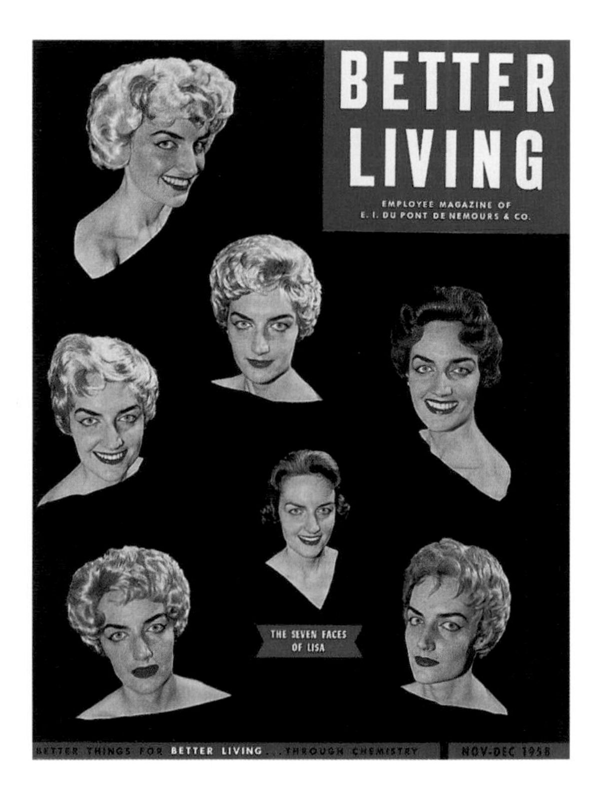

A Love Affair with Science

Science was re-shaping the world and its awesome power was driven home when the atom was split at a Los Alamos laboratory and, as a direct result, 71,000 people died in an instant at Hiroshima. Remarkably, the most astonishing scientific discoveries were virtually invisible – $50 million-worth of the lethal, newly discovered substance plutonium could be held in the bottom of just one test tube – so it is not surprising that the scientist loomed large in the popular culture of the 1950s. Where would scientists lead the world next – into space probably – but during the first few years of peace, the products of the chemical industry were diverted away from bombs and into beauty and domestic products, into dyes, cosmetics, textiles, clothes and home furnishings.

At the time, there was a visible link between the positive image of science and technology and the popular enthusiasm for synthetics. Science, design and popular culture began to merge – molecular prints on fabrics appeared everywhere. In the United States the record 'Molecule a' Go-Go' was released during the twist era, based on a dance invented by a chemistry professor, the movements of which imitated the actions of molecular chemical attraction. The Drifters released 'Nylon Stockings' and 'Nylon' sung by The Nylons was another similar homage to the wonders of the new chemical age. Science was providing labour-saving fashions for all and the white-coated figure (accessorized with test tubes) became a popular hero in a stream of fabric advertisements.

Because fabrics and clothes could be produced more cheaply using synthetics they could also become transient – the essence of fashion. Durability was a tired, pre-war and wartime word. Ultimately, the high turnover of synthetic-based clothing reduced costs and prices and brought a previously unknown range of clothing choice to the mass market. Synthetics had not yet been tainted by a class association – they had no history and so were the perfect choice for the bright new world that would bloom from British and European bomb sites. Nylon, the marvel fibre, 'finer than a spider's thread, stronger than steel and more elegant than silk', was miraculously transformed into an infinite variety of life's minor luxuries. So versatile was the nylon molecule that it was made into the hairspray that supported fashionable towering bouffants; into Christmas trees, luminous fluorescent socks, net crinoline petticoats and artificial mink, Persian Lamb and leopard-skin coats. Its applications seemed limitless. Flying in the face of comfort, nylon was made into drip-dry shirts, men's socks and underwear and blended into suits and linings, all of which sealed in perspiration and kept out oxygen. This lightweight wonder material was particularly recommended for travel to hot climes and somehow its negative malodorous aspects were overlooked. Everyone, it seemed, was enchanted with the novelty and convenience of the new polymers and changing lifestyles and leisure habits seemed only to emphasize the usefulness of synthetics – what could be more appropriate for a skiing holiday than polyester ski pants and a quilted nylon anorak?

Above: Glamour and luxury for all: synthetic fur as illustrated by Du Pont. The *Du Pont Magazine* was a vehicle for stylish photo shoots. Opposite: Wigs of coloured 'Tynex' nylon filaments made by a New York wig manufacturer. More than 5,000 were sold in the summer of 1959 alone, priced at $35 each. A wig of real hair sold for $275.

Feminine Charms and Powers

The 1950s was the decade of the womanly-woman. Emphatically feminine silhouettes were very shapely, tightly fitted over bust and hips or with ballooning crinoline or ball-gown skirts. Synthetic fibres had a large part to play in making real the fantasy of the glamorous female. Wayward figures were controlled by new synthetic-rubber corselets and affordable synthetic fabrics could be bought in the tens of yards to be made up into sweeping romantic gowns. It is widely acknowledged that working women were in need of 're-domestication' at the end of the war and the way in which both governments and the media allied themselves to this aim is well documented – but the side-effect of this was that industry, too, became more home-consumer orientated.

The large number of newly-married couples on both sides of the Atlantic aspired to products that were modern, newly-made and for and by their generation. Out would go inherited and second-hand furniture and in would come contemporary designs made possible because of the fresh new materials. The progressive and overwhelming domestic emphasis of production and consumption gave women an unprecedented degree of aesthetic control. The familiar, stereotypical image of fifties femininity masked a new financial power that was wielded by women in the post-war era. Within their purses, housewives held the fate of some of the world's most powerful manufacturing conglomerates, including Du Pont, ICI and Courtaulds, and this must surely have given women their first taste of real economic power. Communication was the key to manufacturing riches – communication between the corporate chemical giants and the individual woman. Du Pont and its British counterparts well understood the importance of getting their message across and fortunes were spent in competitive advertising.

The desire to inform extended in-house for Du Pont. The pages of their bi-monthly employee magazine *Better Living* reflected the social climate and attitudes of the time. Published between 1946 and 1972, it was in many ways similar in layout and content to its contemporary, *Life* magazine, with virtually every article illustrated by Du Pont's own highly skilled staff photographers. (Du Pont had two impressive vehicles of company publicity, the in-house magazine *Better Living* and the long-established *Du Pont Magazine* which had a worldwide circulation to manufacturers and retailers.) Emphasizing the post-war prosperity that many Americans enjoyed, *Better Living* showed employees how their work led to a better standard of living for others – and particularly for citizens of the United States. Unlike workers in the consumer-goods industries, it was difficult for Du Pont employees to visualize exactly how the chemical industries were contributing to a better life. This was the challenge for the magazine's photojournalists and Alex Henderson took an outstanding series of documentary photographs which were remarkable for their impact. His signature technique was to show multiples of products that were dependent on Du Pont chemicals, illustrating the challenges met by Du Pont's research department, which in turn served to give the workforce a sense of wellbeing, security and stability.

In one revealing article, 'Who Buys the Pants?', an Alex Henderson photograph shows the 600 major clothing items a wife 'would have bought or helped select' for her husband over a twenty-year period'.[23] This image was used to demonstrate the results of a survey carried out for Du Pont in 1959 by a New York market research firm, W. R. Simmons and Associates. Based on the analysis of 1,456 interviews, the researchers were able to conclude that over 85 per cent of wives believed that it was their responsibility to see that their husbands were well-dressed. According to the report, the American male

preferred to surrender all his clothes-buying decisions first to his mother and then to his wife:

'From the moment the first diaper is pinned around his kicking, spidery legs, the American male spends most of his waking and sleeping hours in clothing selected for him by women. Until late in his teens, his mother's purse and prejudices dictate the cut of his coat, the color of his shirts, the weight and weave of his suits, and the hue and length of his hose. Then, for a few brief years of confused freedom, the young man haltingly learns to clothe himself. But the suede servitude of marriage puts a firm stop to this short stint of sartorial license. As he takes a bride, the young man carries across the threshold a costume consultant who will most likely help to mold, manage, and maintain his wardrobe with ever-increasing fervor.'[24]

Colour and pattern were acknowledged to be 'male blind spots' and wives were found to be much more daring and adventurous in the choices that they 'tutored' their husbands to buy. Wives 'hauled home' more than 60 per cent of men's shirts and picked out most husbands' ties and socks but, more than this, the 'impact of the wife's whims and wishes' gradually increases throughout the marriage. On the major purchases such as suits and topcoats, wives were present 47 per cent of the time and exerted a strong influence on the purchase of sportswear, going 'all out for gay sport togs, often plugging styles which men still look on with alarm'. (Husbands were mostly left free to buy hats and shoes on their own.)

This information was of more than passing interest to Du Pont for reasons that the research team made clear: 'Since wives in general showed more knowledge of synthetic fibers and 60 per cent indicated greater desire for fabrics made of these fibers, the advantage to Du Pont of the wife's influence is soon apparent.... In highly competitive textile markets (accounting [in 1959] for over 25 per cent of Du Pont's business), Du Pont fibers contest keenly for consumer favor. The survey makes clear that clothes designed for men must do more than please the man who wears the pants; they much appeal also to the style-conscious wife who goes along and helps her husband make his choice.'[25]

Simultaneously patronized and revered for their buying power and purchasing influence, women were, at least, in the limelight. Fitting titles were handed out: Du Pont's First Lady of Poultry celebrated the

Above: To celebrate the benefits of a new acrylic fibre, the 'Orlon' Queen is chosen at Waynesbro, Virginia in 1949.
Opposite: 'Who Buys the Pants?' – the wife. A Du Pont photograph showing the 600 major clothing items a wife 'would have bought or helped select for her husband' during 20 years of marriage.

benefits of chemical farming and the Orlon Queen beauty contests of 1949 drew attention to the high-fashion allure of new acrylic yarns. Du Pont recorded many salient points about women's purchasing priorities gleaned from its extensive consumer research in an article entitled 'How you Help to Tailor Textiles'. Based on a study of nightwear and lingerie, price lagged behind ease of care, the brand name and comfort and style topped them all. 'We use clothing to communicate with other people – to convey who and what we are. The ivy league tailored executive and the polo-coated teen-age girl alike are creating an image of themselves.... Researchers thus indicate that every step should be taken to make dress-buying a "satisfying experience" – to give the shopper plenty of information. Performance of fibre and ease of care should be fully described, for as one research organization concluded: "Many women are able to use performance features as a personal excuse to buy a dress they really want for psychological reasons".'[26]

While much of Du Pont's market probing was done by in-house teams, the bigger cross-country surveys were farmed out to professional research organizations. Every programme was conducted with the most meticulous care; goals were defined, hypotheses formed, theories tested and every effort was made to ensure that the survey was objective and

bias free. Garments made from Du Pont textiles carried questionnaire postcards but there were more intensive studies where Americans across the country recorded every daily textile purchase and the price paid. Door-to-door surveys were made of housewives' opinions; telephone interviewing was carried out on attitudes towards television advertising; and surveys were undertaken of store buyers' and retailers' views. All this information went into the planning of production, to inform merchandising strategies, to train sales staff and to conduct consumer education campaigns. Data was also distributed in reports to the trade and through speeches, presentations, trade clinics, conventions, trade shows, news releases, consumer literature and advertisements.

'A Quick Tickle with the Iron'

The greatest fans of wash-and-wear were found to be people under 45 and it was found that 88 per cent of women admired the labour-saving qualities of synthetics and, crucially, 94 per cent specified ironing as the task that most women disliked with the 'pet peeve' being men's shirts.[27] Consequently, much fibre advertising was pointedly directed at the fashion conscious but practical young housewife with the emphasis being on the advantages of easy-care, hygienic and, most importantly, non-iron garments. Any analysis of the advertising images and marketing slogans of this time gives an illuminating insight into the rigid gender roles. New fabrics and even male clothing were promoted and sold to women, on the basis of their ease of laundering and compatibility with the newly available domestic washing machine. These were the ideal fabrics for the busy working woman. The new synthetic garments seemed miraculous – permanently-pleated skirts, wash-and-wear non-iron drip-dry shirts, lightweight, stain-resistant suits, and easy-care, low cost, mass fashion. Not only was fashion 'democratized' by becoming more widely accessible, but clothes entered what appeared to be a new era of technological advancement. Magazine editors urged their readers to throw away their ironing boards and dress themselves

Left, top and bottom: In the 1950s, fibres, fabric manufacturers and fashion designers began to be advertised simultaneously in glamorous campaigns that emphasized femininity and elegance.
Opposite: The brilliant photographer, Erwin Blumenfeld, captured the figure controlling secret behind the shapely silhouettes of the 1950s.

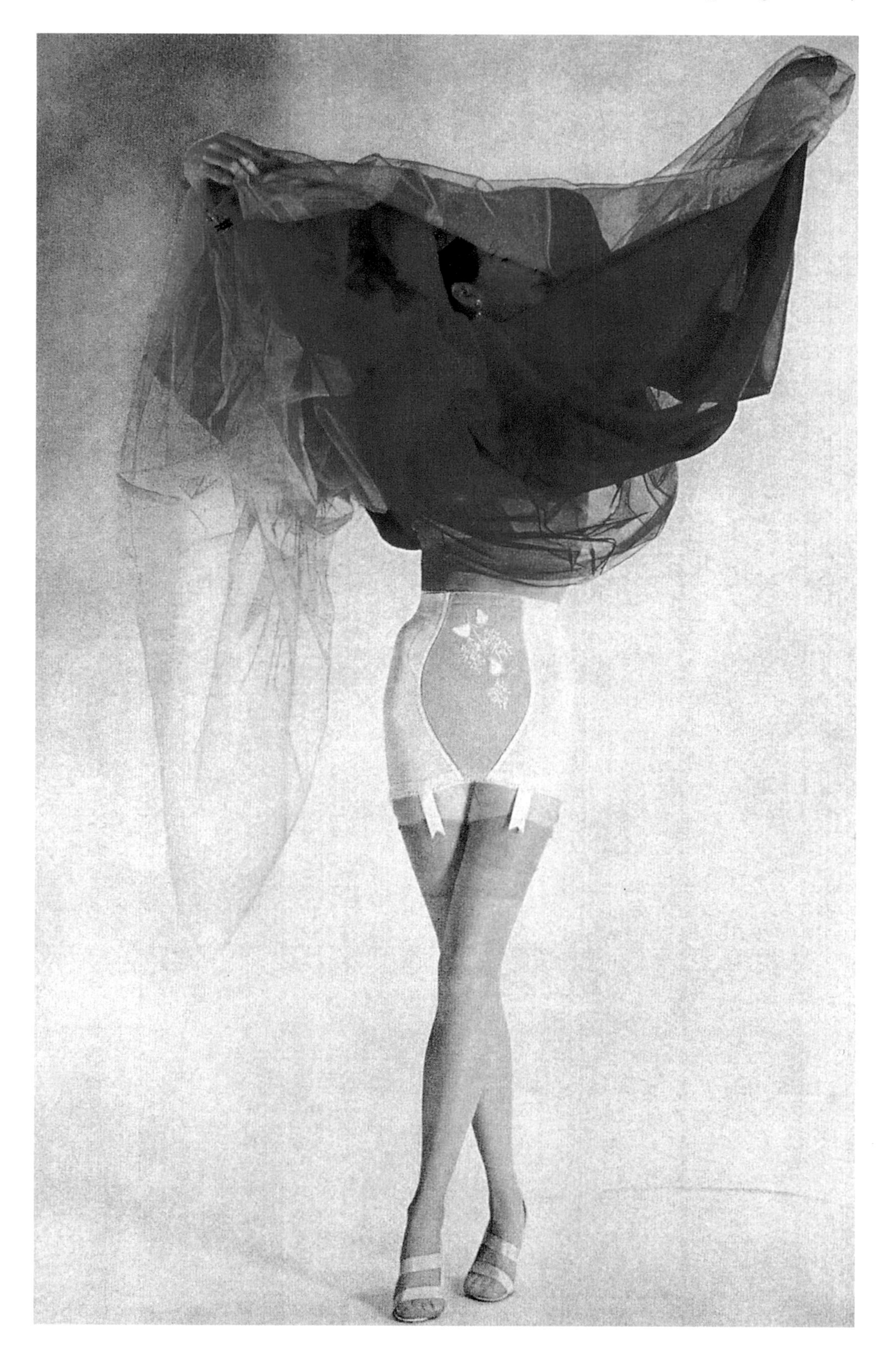

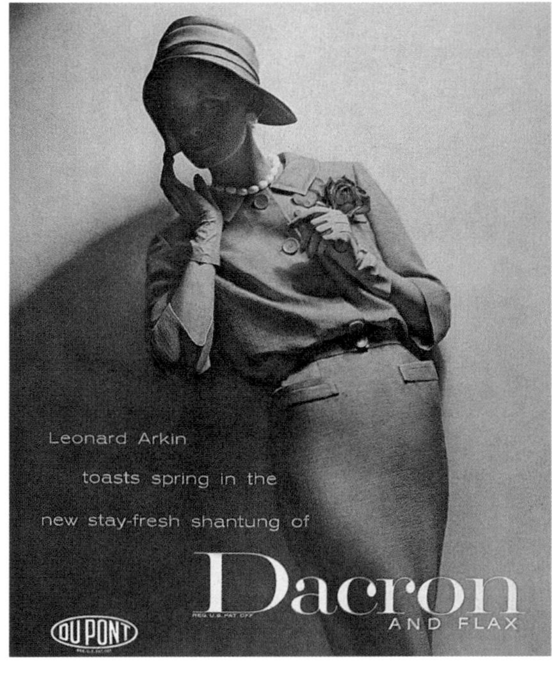

Above: Du Pont show that polyester can be glamorous as well as easy-care. 'Dacron' advertisement.
Opposite, top: 'Orlon', the new synthetic wool that brought a boom in casual clothes and sweater sales.
Opposite, bottom: Monsanto Textile's 'Acrilan', another acrylic wool substitute, was given a go-getting, masculine image.

and their husbands in spellbinding Bri-Nylon – the perfect crease-proof fabric that could be laundered in a newly acquired washing machine.

The drudgery attached to the laundering of garments is of immense significance in women's history. As even the briefest social study will illustrate, laundry is traditionally women's business and natural fibres – and viscose rayon – imposed labour-intensive, time-consuming and exhausting hand-washing and ironing. Consequently, the amount of work involved in cleaning clothes exerted a huge influence on women's choice of fabrics. The arrival of the domestic washing machine went hand-in-hand with the appearance of polyester and nylon garments – a double boon to the post-war working woman. Every woman's holy grail was the drip-dry shirt and any garment that only needed 'a quick tickle with the iron' was invariably an attractive proposition (although considerable care had to be taken to prevent the melting and scorching of early man-made fabrics).

The future was to be man-made and it was only necessary to fill the wardrobe with thermoplastically pleated skirts and blouses and to furnish the home with plastics in virtually every form – Dralon-upholstered furniture, melamine crockery, laminated kitchen surfaces, shiny new radios and televisions, plastic discs and record players and nylon curtains, carpets and sheets. Both Courtaulds and British Celanese marketed 'triacetate', a fibre better known as 'Tricel' which was especially adapted for heat-set permanent pleating and which was also drip-dry and crease resistant. Although acetates were first invented as early as 1914, there were technical problems with the fibres which prevented their use until the 1950s, but Tricel soon became one of the most widely promoted and extensively used of all the acetates.

Beauty is in the Blend

It should be remembered that the vast majority of synthetic fibres were, and still are, most commonly used in blends, what *American Fabrics* described as the great series of 'hybrid fabrics', the 'mating of natural with man-made'. Although polyester was often used in blends with cotton or wool, in the fifties it was proudly promoted as the dominant fibre in men's socks, shirts and in all forms of womenswear, including lingerie. *American Fabrics* reported that blended synthetics have 'provided creative possibilities and new ideas for fashions which had been hitherto undreamed of. They have permitted mills to operate and cutters to make end-products when a rising population plus wartime over-consumption brought about a demand for fabrics of all types far greater than natural-fiber production could handle.'[28]

The economic value of blending fabrics should never be underestimated since the scarcity of wool, for example, put traditional 100 per cent woollen cloths out of the mass-consumer range. Typical of the proportion of these blends of the 1950s were 45 per cent wool to 55 per cent Orlon for men's shirtings, sportswear and women's dress fabrics and 45 per cent worsted and 55 per cent Dacron for suitings. Other more unusual blends included the 'Bunara' dress fabric at 65 per cent wool, 20 per cent rabbit hair and 15 per cent nylon and 'Casmet', a dress and sportswear mix of 70 per cent wool, 15 per cent nylon and 15 per cent fur fibre.

In 1951, *American Fabrics* offered its readers 'the Key to Today's New Combinations of the Three Great Fibers... Nylon, Dacron & Orlon' (all Du Pont products). Their basic properties were defined as follows: Nylon was light, strong and flame-resistant,

Dacron could retain a sharp crease almost indefinitely, had easy spot removal and outstanding wrinkle-recovery and was insensitive to moisture, while Orlon resisted the degradation that came with sunlight, heat, atmospheric gases and perspiration while having a luxurious, warm and dry feel.[29] No longer restricted by the limitations of natural fibres, with the new era of man-mades the horizons of fabric design were greatly expanded. Cloth became a substance that was engineered rather than simply woven. *American Fabrics* described how the 'fabric engineer chooses his combination from a dozen fibers, and, considering the exact end use desired, engineers his construction accordingly'. Examples of every possible synthetic and natural combination were recommended – nylon plus cotton, Orlon, wool, rayon, acetate or silk, or all nylon, the last being 'ideal for travelling because it requires only a minimum of ironing'. A postscript pointed out that all three of these Du Pont fibres were still in short supply but the company was well on the way to remedying this. What this article represented was an advance notice of the coming textile revolution – consumers were being educated and prepared.

Onto the High Street

Nylon, polyester, acetate and acrylic were ultimately the fibres that delivered high fashion to the high street, where chain stores regularly purchased Paris couture gowns for the sole purpose of copying them by the thousand. The newly branded fibres were rapidly taken up by stores and translated into affordable fashion garments, and mail order companies, such as Kays, introduced millions of their customers to brands like Tricel and Crimplene. In Britain, the famous retailers Marks & Spencer epitomized the progress towards mass fashion by incorporating synthetic fabrics into their garments. The company set up its own experimental textile laboratory which tested the new materials and instructed its suppliers and customers alike on wearability, washability, durability and colour-fastness.

Many technological changes were pioneered by the Marks & Spencer knitwear department and as early as 1947 nylon was being introduced to M&S labelled lingerie, stockings, blouses and dresses, and by 1953 one reporter described the contribution of M&S to fashion as ensuring that 'every woman could have some top quality nylon in her wardrobe'. Terylene (the British polyester) brought the first permanently-

Above: Marks and Spencer, the celebrated British retailer, introduced millions of their customers to the new synthetic fibres and brands. The wonder of 'Terylene' permanently-pleated skirts came in 1954.

Opposite: The sweater girl was a new fashion phenomenon in the 1950s, created by the mass availability of lightweight wool substitutes such as 'Orlon'.

pleated skirts to M&S in 1954 and – as proof of the virtual indestructibility of this fibre – many of these sun-ray pleated skirts still survive. Within a year or two, the mass market also met Orlon sweaters, Tricel blouses and Lycra girdles, thanks to Marks & Spencer.

Orlon (originally called 'Fiber A') was first announced in 1949 and was another of Du Pont's 'easy living' fibres – its two outstanding qualities were its resistance to outdoor exposure and acid and a remarkable capacity to form staple fibre, thereby creating the bulkiness of wool. It had all the lightness and durability associated with wool but with an acid-resistant toughness that made it the perfect material for work clothes and uniforms. Orlon uniforms were described in the *Du Pont Magazine* as early as 1952, an end use that was also rightly predicted for Du Pont's polyester. Wool was famous for its washing difficulties; being prone to felting, it had to be hand washed, squeezed in towels, re-shaped and dried flat, whereas Orlon (a plastic form of wool) was characterized by easy-washing fast-drying qualities. It was crease-, moth-, mildew- and perspiration-resistant and considered a fashionable fibre for sportswear, dresses, coats, skirts, separates, suits, childrenswear and negligés.

In the main, acrylics replaced wool as the basic sweater material and brought about an enormous expansion in the fashion knitwear industry. Orlon looked like wool, it could be fluffy or fitted, but it differed in its weight and colourways. Where wool was heavy, acrylic was light, and where the palette of wool was muted, acrylic could be pure white or vividly dyed. Natural wool was expensive in itself and expensive to clean, involving careful washing, blocking and pressing while acrylics were inexpensive and easy-care. Most of all, the threat of moths could be obliterated with acrylics – insects had no appetite for synthetics. In 1937, Lana Turner was declared the first American Sweater Girl and during the 1950s figure-hugging knits became a fashion success story. Britain produced its own home-grown Sweater Girl, Jayne Mansfield. By 1959, Du Pont declared that 'the sweater industry and the fashion world were launched in the beginning of a stylish revolution' and that sales had doubled in just seven years, with 100 million sweaters being sold each year (two-thirds of which were made of Orlon or Du Pont nylon).[30]

Orlon brought the year-round sweater into existence and changed it from a strictly utilitarian item to an important classic in the fashion wardrobe. Nothing could be more evocative of the 1950s than the twin-set and pearls, both of which essentials were often created from plastics. The dramatic acceptance of Orlon in sweaters is demonstrated by the fact that the fibre represented 5 per cent of the American market in 1953 and 50 per cent in 1960 – a statistic that was alarming to the international wool industry.[31]

The Brand and Image War

Polyesters, nylons, acrylics and acetates enjoyed a huge retail success under a multitude of fifties brand names. Polyester fibres included Dacron, Terylene, Trelenka and Crimplene; acrylic wool substitutes included Du Pont's Orlon, Courtaulds' Courtelle and Monsanto Textile's Acrilan; and the main acetate fibre then available was British Celanese's Tricel. Marketing strategists clung to the laboratory-sounding names that befitted the atomic age which had brought a whole new scientific vocabulary (neutrons, hydrogen, plutonium and so on) into everyday speech. The sometimes baffling array of brand names was intended to generate the impression of uniqueness, of a mysterious chemical formula especially made for that drip-dry shirt or polyester sun-ray skirt.

The traditional fibre industries of wool, silk and cotton came under siege from synthetics and a symptom of their economic apprehension was the profusion of promotional advertisements they placed in both trade and fashion magazines. The competition between natural and synthetic fibre producers must have instigated one of the most profitable eras for magazines whose pages became the battleground for fashion credibility. The British Rayon and Synthetic Fibres Federation moved into smart new London headquarters in Hamilton House, Piccadilly in 1953 and celebrated the occasion with an exhibition and fashion show featuring rayon and rayon mix fabrics. Famous designers, all members of the Incorporated Society of London Fashion Designers, were commissioned to create a collection in rayon, including Hardy Amies, Norman Hartnell, Victor Stiebel and Michael at Lachasse. The story was covered by *The Ambassador* magazine, which in previous issues had featured many of the rayon 'couture' fabrics in question, several of which were produced by British Celanese and Courtaulds.[32]

Provoked by the expansion of all man-made fibres in the fifties and sixties, both cotton and wool manufacturers formed themselves into world trade boards, each with a distinctive logo: the 'pure cotton' flower and the specially designed op art 'woolmark' of spiralled yarns. Pooling resources, these boards also began to invest in research and development that would give the natural fibres some of the wash-and-wear qualities of synthetics. Not only were the natural fibres fighting back but the numerous synthetic brands were also at war with one another, and

everyone wanted to associate their product with a prestigious fashion designer or with a media celebrity. The chemical-fibre producers spent fortunes on this type of advertising, even though they did not manufacture the end products. In 1959, only 5 per cent of Du Pont's products went directly to the retail counter, the rest, including all of its textile fibres, went to other manufacturers for processing and finishing. The marketing of Orlon is a good example of the extraordinary efforts Du Pont made to let the public know about its products and, more importantly, about its customers' products. 'This year [1959], Du Pont is going directly to the public with one of the largest advertising budgets ever allotted to sweaters. Sweaters made of Du Pont's fibers will be advertised in top magazines, including leading fashion magazines and *Life*, *TV Guide*, *This Week* and others. Also, sweaters from Du Pont's fibers will be featured on the Steve Allen TV show, the Dave Garroway show and the Du Pont *Show of the Month*. All of this effort emphasizes Du Pont's continuing belief that the best way it can promote business is to expand older markets and develop new markets for its customers.'[33]

In 1952, Du Pont launched its Product Information section within its Public Relations Department. This new division was to play a key role in the introduction of synthetic materials to the high fashion industry and its primary purpose was to 'create news releases, accompanied by photographs, which would be run

editorially by trade journals and newspapers, in effect creating inexpensive publicity and indirect advertising.... To be used, the news releases had to include genuinely interesting information about a Du Pont product, its development, manufacture, or applications. They could not be direct commercial promotions of a product itself, that being the proper work of the Advertising Department.... Product Information for Textile Fibers was handled out of that department's offices in New York City's Empire State Building. [Their photographs] were issued with press releases for editorial use, designed to establish and strengthen in the reader's mind the connection between Du Pont, maker of fibers, and the garment and fashion industry.'[34]

Du Pont's great rival producers of man-made fibres, the British companies ICI and Courtaulds, followed similar corporate patterns to the American giant; both invested huge capital sums in research and development and in follow-up marketing, promotional and information campaigns, directed at textile and apparel manufacturers as well as at consumers. The new profession of Marketing was then taking hold in Britain, based on techniques that were first developed in and imported from the United States. Consumer research, mass marketing, the need for fibre information and the active use of product branding and psychological advertising completely transformed the outdated British textiles industry. No longer would it be enough to rest on its laurels and relate primarily to clothing manufacturers; from now on sales would have to be consumer led and every effort would have to be made to persuade consumers that one brand name was more desirable than another.

In 1958, British Nylon Spinners introduced its own brand name, 'Bri-Nylon', accompanied by an epidemic of advertising in the press, department stores, magazines and on television, and fashion editorials, textile reports and fashion shows – saturation publicity that praised the advent of the new man-made fashion epoch. Fashion appeal had somehow to be grafted onto the new fabrics – nylon, polyester, acrylic and acetates – and all the fibre companies sponsored major designers to work with their synthetic materials, so that the resulting collections of sample clothes could then be displayed at highly publicized fashion shows. Throughout the 1950s, under the BNS corporate banner, huge nylon fashion shows were held at the Royal Albert Hall in

Opposite: Lady Ashton (Madge Garland), former fashion editor of British *Vogue* and first Professor of Fashion at London's RCA, instructs students in the skills of fabric selection and design in the mid-1950s.

Above: At the first synthetic-fibre fair to be held in Europe, Textil '62 in London, Princess Sibylla of Sweden (centre with flowers) admired evening gowns of Terylene in Terylene Hall.

London to promote the delights of nylon. A great deal of money was spent sponsoring nylon collections from both French and British couturiers, including the Frank Usher label and the Sekers textile company.[35]

London, it seems, was brimming with publicity promoting the newly branded synthetics. Lister was boosting its range of fabrics in Orlon; Littlewoods chain stores held fashion shows celebrating Acrilan and Bri-Nylon; Celanese House and the British Nylon Spinners' showroom at Bowater House in Knightsbridge both provided space to give man-mades a fashion profile. Their motive was to connect nylon with the high fashion market and these prestigious shows were staged to show that highly esteemed designers like Hardy Amies were prepared to sanction nylon for couture standard gowns. The essential objective of ICI and British Nylon Spinners was to give their intangible product some form of identifiable image in garment form and to stimulate brand loyalty by advertising nationally and internationally. Garment manufacturers and designers using Bri-Nylon, Du Pont nylon or any other major fibre benefited from a mass of free advertising simply because the chemical companies wanted to sell an invisible product by associating it with a highly familiar one.

In addition to this, the proven and favoured system of expert or celebrity endorsement was freely exercised. The former fashion editor of British *Vogue*, Lady Ashton (Madge Garland), described for ICI the wonders of Terylene in a 1954 issue of *Queen*. In her then role as Professor of Fashion at the Royal College of Art in London, she declared that Terylene was the 'latest, most dramatic' of synthetic fibres, which would bring 'untold blessings to us all'. Potential customers were being prepared for the soon-to-be-available new material and, recognizing their need to be educated as to its chemistry and its benefits, Lady Ashton continued, '"Terylene" has two distinct forms. One is called staple fibre and the other filament yarn. Between them they make (or will soon make) materials for practically all the clothes you wear. Your suits, skirts and winter woollies will be made of staple fibre fabrics; your flouncy ball dresses, fine undies and "silky" frocks will be of filament yarn.'[36] In later issues her readers were reassured that, although 'The word synthetic is apt to suggest something rather cold and chemical, let me quickly say that one of the joys of "Terylene" is its warm, friendly, "natural" feel. A quality unique in synthetics in my experience.... "Terylene" is likely to affect your life. [The] fabrics are remarkably resistant to wrinkling, creasing, bagging and sagging. They are just as strong wet as dry. What a boon "Terylene" will be to women who hate mending (and who doesn't?)

..."Terylene will bring you a variety of lovely long-wearing clothes – from swim-suits to twin-sets, from corsets to coats...and will begin to enter into – and glamorise – every aspect of your life.'[37]

Both Terylene and Dacron were polyesters and the only difference was that the former was made by ICI and the latter by Du Pont. Both materials were portrayed as representing the liberated, practical, modern woman. A Dacron shirt dress would 'look starched without any starch and stay that way. Let it get rained on, or caught in a fog, and in no time it will bounce right back to its pristine freshness.'[38] Competitive branding was so widespread that it is hardly surprising that consumer confusion was rife during the fifties and sixties. Textile manufacturers insisted on identifying each of their fibres with a different trademark so that consumers often 'did not know whether an identifying name represented a particular kind of fibre or a trademark for some kind of newly created fibre.'[39] There were hundreds of synthetic trademarks, including Acele, Du Pont's acetate, the famous Banlon, a nylon manufactured by Bancroft, and Crimplene, ICI's infamous polyester.

Courtaulds Expands

As the impact of full synthetic fibres began to be realized, the future of rayon came into question. Of the British rayon makers, only Courtaulds was avoiding financial difficulties by improving the quality

and versatility of its rayon staple and by developing a triacetate yarn and an acrylic fibre – 'Tricel' and 'Courtelle' (the first British acrylic fibre) – both of which went into production in the mid-1950s. Courtaulds' policy of diversification protected it from the worst effects of the shift to synthetics. Between 1957 and 1961, Courtaulds acquired several other companies valued at over £44 million.[40] The largest of these was British Celanese, an early rayon producer whose profits had been diminishing steadily. Five smaller rayon companies were also absorbed. Although Courtaulds had become involved in synthetic fibres rather late in the day compared with Du Pont, the sales of Tricel and Courtelle increased company profits to a record level between 1959 and 1960 – just as the company was about to experience one of its most troubled times. There was intense competition between Courtaulds and ICI and Courtaulds' systematic acquisition of companies that already manufactured both textiles and clothing (such as British Celanese, which had its own knitting and garment-making subsidiaries) was part of its goal to increase market share.The relationship between the BNS partners was sometimes less than harmonious until, in December 1961, ICI made a takeover bid for Courtaulds. A public battle ensued and, as this was the largest take-over bid in British industrial history to date, the conflict attracted a great deal of press attention and even provoked a debate in Parliament. The government decided not to intervene but by 12 March 1962, ICI had to concede defeat. After this, Courtaulds moved even further into the synthetic textile industry with the production of its nylon 6, marketed as 'Celon' during the early 1960s.

During the three decades following the war, Courtaulds invested heavily in all three of its main synthetics, nylon, acrylic and polyester; consequently, it needed to secure textile and clothing outlets for its textile fibres. It was Courtaulds' policy to develop 'downsteam' in the textile industry by taking over specialist firms at all levels of textile manufacture and marketing, including dyers, weavers, fabric printers and wholesalers. After the wave of takeovers in the late fifties and early sixties, which had included the lingerie manufacturer Gossard in 1957, acquisitions continued with the purchase of the womenswear companies Meridian and Susan Small (1963), followed by the stocking manufacturers Aristock, Kayser, Bondor and Wolsey. Between 1962 and 1969,

Courtaulds invested more than £175 million in acquisitions and £227 million in upgrading equipment and in expansion – mostly into fibres and textiles. Courtaulds thrived on the booming popularity of knitted jersey garments, and its financial success was furthered by the rapid growth in the buying power of the big retail chain stores such as Marks and Spencer.

Branding Hysteria

Because there were really very few generic textile fibres – viscose, acetate, polyamides and polymers – the key to the manufacturers' success lay in brand differentiation. In this hugely competitive industry, patenting was the crucial way to establish an individual identity that could win the loyalty of consumers. Product branding was the main weapon and, although consumers may have been buying, time and again, exactly the same fibre (made by just a handful of petrochemical giants), they had the illusion of luxuriating in an ever expanding abundance of fabric choice. Courtaulds, Du Pont and ICI, alongside the other giant petrochemical firms like Dow, Union Carbide and Celanese in the United States and Rhone Poulenc, Montedison and IG Farben in Europe, and yet others in Japan, have continued to market an infinite number of branded fibres since the 1950s. By 1969, Du Pont had invented 31 polyesters and 70 nylons; in 1971 they were producing 71 polyesters and today there are hundreds of different types.

Synthetics created a very visible increase in fashion advertising during the 1950s which increased public awareness of designers and changing fashion silhouettes and stimulated an interest in new textiles. The marketing frenzy that so characterized the decade or so after the war was centred on making the few new synthetic fibres familiar under the guise of branded textile products. Nylon was identified as Bri-Nylon or Celon, polyester as Terylene, Dacron or Crimplene, and acrylic as Courtelle Orlon or Acrilan, among many others. The first significant impact was the way these textiles made fashionable clothes more available to a mass market – although, interestingly, these new materials were made into conventional garment forms. Apart from their capacity for permanent pleating and the easy-care benefits, synthetics had no specific design influence that was based on their inherent characteristics. The reason for this is that the primary intention of the fibre manufacturers was (and continues to be) to sell the maximum amount

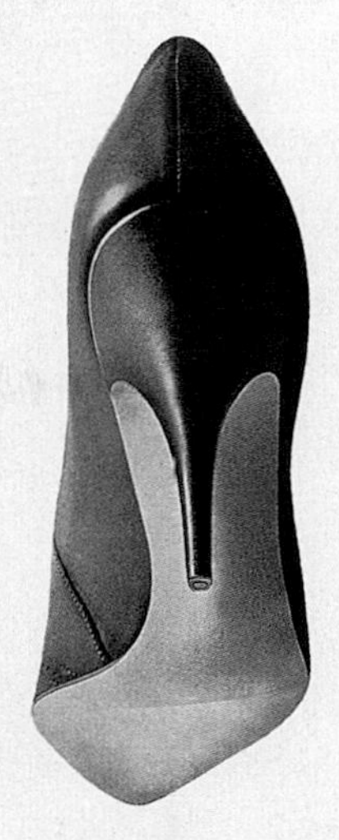
as every
stiletto
knows
BRI-NYLON KEEPS A CARPET YOUNGER
BRI-NYLON CARPET FIBRE—specially made for blending into carpets—keeps your carpet looking fresh while other carpets age.
SEE THE NAME BRI-NYLON is on the label or pattern book before you buy your carpet. Most famous manufacturers use it.

BRI*
NYLON
Wool/BRI-NYLON blended carpets are being widely produced. Some of the well-known manufacturers who make them are:
BRINTONS, B.C.L., CARPET TRADES, CARPET MANUFACTURING CO., FIRTHS, GRAY'S OF AYR, KOSSET, T. & A. NAYLOR, WILTON ROYAL, SOLENT, WOODWARD GROSVENOR.

Above: By the early 1960s, acrylic and nylcn were replacing wool in the home as well as in the wardrobe. Bri-Nylon carpets were tough enough to withstand fashionable stilettos. Advertisement in *The Ambassador* magazine, 1963.
Opposite: *Six Faces of Terylene* was a film made by ICI in 1964 to show the versatility of the fabric for the modern woman.

of their product and, in their view, the best way to do that is to appeal directly to the mass market, thereby building loyalty to the fibre, not to the designer.

It was not until the early 1960s that the real fashion potential of synthetics began to be unlocked and exploited, when designers themselves were looking for new materials and fabrics with which to undermine the conventions of traditional couture. Before then, synthetic fabrics had simply meant more of the same, but the sixties introduced more and different – silhouettes, cuts, textures and colours that had never been seen before, all of which depended on the inherent chemical nature of synthetic fabrics. The sixties and early seventies were to be the heyday for synthetic fashions, when design at last took centre stage and the shock value of the new materials began to be thoroughly enjoyed.

4

> 'Don't try to sell the steak sell the sizzle! A steak in the abstract is attractive enough, but it takes the sizzle and the aroma to make the taste buds cry out for satisfaction. In the textile business, a fiber can be compared to Wheeler's steak. No cones of yarn, fresh from the spinning machine, win squeals of admiration in Paris or on Fifth Avenue. It's the dress cut from the fabric woven from the yarn that puts the sizzle in the salons.'
>
> Elmer Wheeler, 'Super Salesman', 1955

1955–70

Paris Couture Embraces Man-Mades

Traditionally, Parisian couture was the shop window for French luxury fabrics, those made only from the highest quality 'noble' natural fibres, usually silks. Many of the great French textile companies have financed couture collections as loss leaders in order to showcase their fabrics. Perhaps the most famous example of a textile company funding a designer is when Marcel Boussac funded Christian Dior's debut and sensational 'New Look' collection in 1947. After the rationing and austerity of war, the sheer luxuriousness and femininity of Dior's designs reclaimed the title of world fashion capital for Paris.

At the heart of the New Look was the huge quantity of fabric needed to create the long, bias-cut, circular skirts; around 13.5 metres (15 yards) of fabric were used in a single cocktail dress. In textile terms it was like a return to the Victorian crinoline and manufacturers were delighted by the increased demand for fashion fabrics. Dior's success was an almost freakish consequence of the fashion deprivations of wartime; he simply gave women what they had been denied and yearned for – exaggerated curves and an excess of fabrics made into totally impractical clothes. What Dior did not give to fashion was a 'new' look – it was, in fact, a very old one, a mixture of late nineteenth-century silhouettes which brought to the fifties a renewed romantic ideal of curvaceous femininity.

Dior was the declared King of Paris couture and, for a time, the hierarchical system of fashion seemed to have been restored intact, but within just a few years, many of the post-war social and cultural shifts

Opposite: A Du Pont nylon tulle ballgown designed by the couturier Madame Grès in 1961.

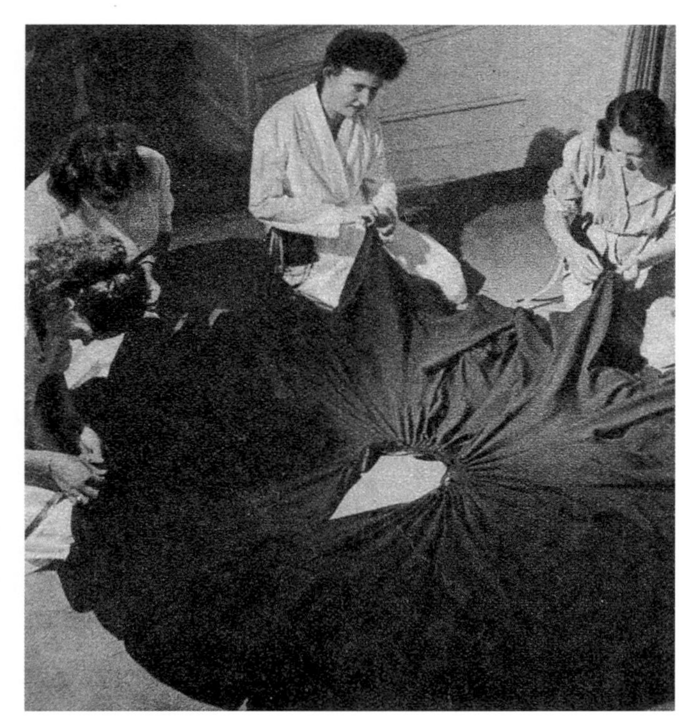

Above: 'Paris Forgets it's 1947' headlined Britain's *Picture Post* when Dior launched his New Look. Ignoring post-war fabric shortages and clothes rationing, salon seamstresses work on a wildly extravagant circular skirt.
Below: Christian Dior's dinner suit in ruby silk velvet epitomizes the luxurious elegance of Paris fashion. *Harper's Bazaar*, 1949.

began to make their mark, even in the fragrant salons of the Rue de la Paix and the Avenue Montaigne. Couture was in financial crisis and was increasingly out of step with its times. The couture system had to undergo a complete metamorphosis in the changed economic and social world of post-war Europe. Its wealthy client-base had virtually disappeared and its survival came to depend more and more on the licensing of perfumes and all the other small luxury items made for a mass market. The designer was to become a brand name and, as a consequence, the function of the couture fashion show shifted from that of selling actual garments to actual clients to the modern purpose of generating the vital publicity needed to promote the designer as a global trademark. Pierre Cardin was the most entrepreneurial of all designers, 'the man who became a label', who, from the 1960s, put his name to thousands of licenced products, including ties, cigarettes, frying pans, chocolates, alarm clocks and aeroplanes.[1] The 1950s marked the beginning of the modern fashion era, the beginning of real fashion democratization – and the beginning of the end of the exclusivity perpetuated and jealously protected by the Chambre Syndicale in Paris, the organization founded in 1868 that represents the French couture industry. Within a decade, the one-off couture gown would be little more than a symbol and the orientation of couture would be moving ever closer to ready-to-wear, via the 'boutique' collections of the Left Bank.

The sixties became a spectacularly modish era for all technological materials and, in particular, a boom period for synthetic fashions. The singular fact about this phenomenon is that synthetics were considered avant-garde by both ends of the fashion spectrum – by Parisian couturiers and by London's young, anti-establishment boutique owners. Ironically, synthetics both modernized the image of the one and made possible the fun, throwaway clothes created by the other. The next chapter will take up the story in London but first the conquest of Paris by the makers of synthetic materials will be explored.

The King can do no Wrong

The fading away of its traditional clients, the trend towards technology and the ever improving ready-to-wear industry combined to make times uncertain for haute couture designers. This was the perfect moment for synthetic-fibre producers to court both Parisian and the newly emerging American couturiers. Du Pont seized this opportunity to win over the top fashion names to their branded fibres, to Dacron, Orlon and Antron (a nylon), and made strenuous and expensive endeavours to relate to the fickle world of high fashion. From its earliest ventures into fashion fibres, Du Pont was acutely conscious of the marketing significance of creating the all-important high-status link with Parisian couture. In 1927, for example (when the company was first promoting its rayon fibres), a lengthy marketing report described the deep psychological significance of a high fashion link – true then, true in the sixties and still true today: 'We plan to secure for rayon the endorsements of the most famous Parisian costume designers – such as Paquin, Patou, Lelong, Poiret, Chanel, Boulanger, Vionnet and others of a similar standing and prestige. We will send to Paris a representative who will interview these famous couturiers and induce them either to make special creations in rayon or to interpret their current creations in rayon. We will also induce them to issue a signed statement expressing their approval of rayon as a fabric suited to the requirements of the mode.' The importance of such endorsements and of their benefit to Du Pont's fibres 'could hardly be over-estimated'. The great couturiers 'speak with the voice of authority. No matter how prejudiced a woman may be against rayon, that prejudice will not survive an endorsement by a Parisian designer. It is the old story that "the king can do no wrong"... A Poiret evening gown of rayon, for instance, builds far more rayon prestige than rayon drapery, rayon hosiery or even rayon underwear.... once rayon has been established as the choice of the Parisian designer, it will readily and speedily be accepted in all its other fields.'[2]

Above: Monsieur Balmain and a Du Pont executive in 1956 examining a luxurious ballgown and wrap made from synthetic fabrics.
Below: Du Pont's nylon fibre was propelled straight towards the heart of fashion – Paris. For many years the company covered the couture collections and publicized the many 'high-fashion creations' made from Du Pont fibres. *Du Pont Magazine*, 1952.

The Making of Desirable Synthetics

From the beginning of its association with textile fibres, Du Pont had to work at improving the design image of its fibres in the eyes of both fabric manufacturers and the end users, the fashion industry. Soon after the first Du Pont rayon plant was opened in Buffalo in 1921, it became clear that the future of this pioneering man-made fibre depended upon making better fabrics from it. Textile mills, unused to working with rayon, had problems manufacturing fabrics from the new fibres so, in response to this, Du Pont employed a Russian émigré, Alexis Sommaripa, to launch their Fabric Development programme in

1926. His idea was that mills should be shown varied and versatile sample fabrics created by Du Pont themselves. To achieve this he made the first of many trips to Europe to bring back trunk loads of silk fabrics gathered from all over the Continent. Using these silks as his starting point he 'hired technical men to help analyze those he was most interested in. Soon hundreds of reproductions of the fabrics, woven from Du Pont rayon, were being shown to customers as examples of what could be done with a man-made fiber. Market after market was won for rayon by this show-me technique and by steady improvement in the fiber itself. Thus the fabric development principle proved itself.'[3] As time passed, Du Pont took an increasingly proactive role in fabric promotion and in textile design education within the trade. By 1959, the Fabric Development department, located in New York's Empire State Building, had a collection of more than 100,000 examples of natural and synthetic textiles, most from Europe but some from China, Japan and South America. These materials were intended to be used as inspiration for new fabrics, for weaves, colours and print designs, and could be viewed by appointment by any of Du Pont's clients.[4]

Textile designing is the most future-orientated of all the trades connected with fashion: while fashion designers may be thinking a year or 18 months ahead, textile designers have to plan some two years or more in advance. With the coming of synthetic fibres, the textile industry became more complicated and the reliance mills had once placed on standard long-run fabrics began to crumble. Synthetics raised consumer expectations for more variety in pattern and texture. Where mills had traditionally produced three or four standard fabrics, one year printed, the next dyed a solid colour, by the mid-1950s they turned out 20 or more different fabrics with a life of seldom more than three years: 'the first it's on the way in, the second it has arrived, and the third it's on the way out'.[5] As the pace of fabric change accelerated, the mills became acutely aware of the need to be ahead of fashion trends. Timing was vital: 'a good fabric offered to garment cutters too early, or too late, can fail as utterly as one completely off the beam'.

Although Du Pont manufactured no dresses, stockings or fabrics commercially, as a service to its mill customers it employed designers to create sample or 'idea' textiles with the retail market in mind.[6] By the mid-1950s, the group designed, produced, dyed and finished close to 1,400 different fabrics a year, most of which were intended for apparel. Sometimes a Du Pont-designed fabric was adopted outright by a mill but more often it was modified. Among the most successful fabrics to come out of the Fabric Development department was 'French crepe', a combination of rayon and acetate, the rayon yarn being given an additional twist for the creping effect. Hundreds of thousands of yards of this fabric went into mass produced women's blouses and lingerie over an eight year period. Wash-and-wear blended fabrics for summer suits also originated with Du Pont, as did a very successful satin of Orlon and silk which was introduced in both the United States and Europe.[7]

Jacques Heim ballgown, 1955. Satin of 100 per cent Orlon.

Attracting and Promoting Haute Couture

Synthetic-fibre companies are in a difficult position; they must prove and demonstrate the desirability of a product that in itself is visually unexciting and this is why their advertising and marketing strategies are so important. Du Pont was in the vanguard from the earliest days. Elmer Wheeler, 'Super Salesman', became an adviser to Du Pont in the mid-fifties and his

Top:
Lanvin's Goya-inspired wedding dress in Orlon satin, 1956.
Jean Patou, Orlon satin with silver and crystal embroidery, 1956.
Pierre Balmain's stylish coat in Orlon, 1955.
Centre:
Jean Patou, suit in a ribbon weave of nylon and rayon, 1957.
Givenchy resort collection in acrylic, acetate and polyester blends, 1958.
Jean Dessès, Dacron and silk cocktail coat with mink trim, 1957.
Bottom:
Dior nylon tulle ballgown, 1956.
Balmain, nylon tulle skirt with Abyssinian panther bodice, 1961.

'Don't Sell The Steak, Sell The Sizzle' slogan was just the right recommendation for its Fabric Development department.[8] The most sizzling gowns came from Paris and, by 1955, *Du Pont Magazine* noted: 'The attitude of the haute couture in France and elsewhere toward man-made fibres has lost the hauteur of a few years back, when one designer disdained nylon on the grounds that it was "a stocking fiber". This year at least 14 fabrics containing Du Pont fibres, including nylon, will be represented in the latest Paris fashion. When European designers were shown fabrics that were aesthetically pleasing and at the same time had new functional virtues, they used them.'[9]

Probably as a consequence of the fall from status of synthetics during the 1970s, a widespread myth has grown up that couturiers *never* used the early nylons, acrylics or acetates. It is surprising how commonplace this opinion is, particularly in Britain where some curators of dress collections and fashion historians simply do not accept that these synthetics were considered stylish enough to be used by 1950s haute couture. There is abundant evidence to the contrary in Du Pont's photographic archive, a small sample of images from 1955–8 showing several Dior nylon gowns and Balmain day and evening dresses in Orlon; Jacques Heim, Lanvin and Patou models all using Orlon acrylic lustrous satins; a Jean Dessès Dacron cocktail coat, a Jacques Griffe suit in a blend of Orlon, wool, mohair and rabbit hair and a Jean Patou suit in nylon and rayon; and a collection of resort clothes by Givenchy in polyester, acrylic and acetate. The most likely explanation for so many examples is that, like all designers, Parisian couturiers were excited and interested in new fabrics and wanted to experiment with them. By the end of the 1950s, Du Pont held a large collection of Parisian and American designer clothing, all made from Du Pont's man-made fibres. Valued at $30,000, these garments were intended to inspire Du Pont's customers.

By the mid-fifties, Du Pont was monitoring Paris couturiers' use of their fibres and one report in 1954 contained some very promising information: 'Because Paris remains the style capital of the world, the increasing use of these fabrics by some of the biggest names in fashion is news.... Nylon was used by Christian Dior, world-renowned for his daring and originality...[while] Hubert de Givenchy...used "Orlon" acrylic fiber and silk in three widely acclaimed dresses. Chanel, whose first collection in 16 years was

Above: *Chemical Week Forecast* sponsored an 'Orlon in Paris' fashion show which brought couture originals to New York in 1955.
Opposite, top: Two New York fashion models in Orlon satin. Du Pont's advertising agency commissioned top photographers like Horst P. Horst to create high class fashion images which were used to promote synthetic fibres.
Opposite, bottom: In 1954 Chanel's first collection for 16 years made outstanding use of nylon, including this afternoon dress in nylon plissé.

the talk of Paris, also made outstanding use of nylon. Nylon tulle, always popular, was shown in bouffant evening gowns. Chiffon-weight nylon jersey...[and] nylon taffetas were prominent...and many mixes of nylon and other fibres. Rich-textured rayon fabrics provided interesting coat news. Some of the most exquisite fabrics of the collections were in magnificent after-five costumes. Here again, rayon and acetate were prominent in rich, heavy fabrics of great elegance....I was delighted to find chemical fibres so extensively used by the Parisians.'[10]

Pierre Balmain's gowns were included in *Chemical Week Forecast*'s fashion show 'Orlon in Paris' which was shown in New York in 1955. In 1956, a *Better Living* cover showed two glamorous New York models who had appeared frequently in Du Pont advertisements. 'Top Fashion Models Help Du Pont Sell' announced the headline and the story inside described how 'a host of America's best-known faces' were promoting man-made fibres. 'Paris Comes to Kinston' was another feature published in *Better Living* in 1958, describing a gala fashion show put on for Du Pont employees and local Kinston people. In celebration of Dacron's fifth anniversary, 19 designs in polyester and other man-made fibres were selected and flown over from 'the Paris salons of famed couturiers like Christian Dior and Jacques Heim' to delight the audience in the home of Dacron manufacture.[11] 'These originals were chosen by Du Pont fashion experts from scores of gowns created this year in the leading salons of Paris', ran the story, 'they were selected...because they incorporate Du Pont's man-made fibres in new and distinctive fabrics. Never had man-made fibres been used as much by the world style-setters as in this year's much-discussed collection' concluded the article. The excitement did not stop at Kinston because the show was then sent on tour to a dozen American metropolitan centres for the benefit of designers and mass-market manufacturers alike. In many instances, American manufacturers produced line-for-line copies of the French originals. These couture designs were also used to help sell patterns for the American dressmaking industry, which closely followed styles introduced in Paris. In Britain, too, ICI were pleased to advertise in 1961 that Givenchy, Cardin, Heim, Nina Ricci, Griffe and Jean Dessès were designing with Terylene.[12]

Everything was being done by Du Pont's Public Relations department to inform the media of the fashion appeal of synthetic fabrics, including the commissioning of top fashion photographers like Horst P. Horst to photograph Dacron fashions. In the 1950s and 1960s a flood of top quality photographs and press releases were circulated all across America by Du Pont, showing and describing the many examples of couture-level designs that used their branded synthetics. Other couture feathers were added to Du Pont's cap in the sixties with

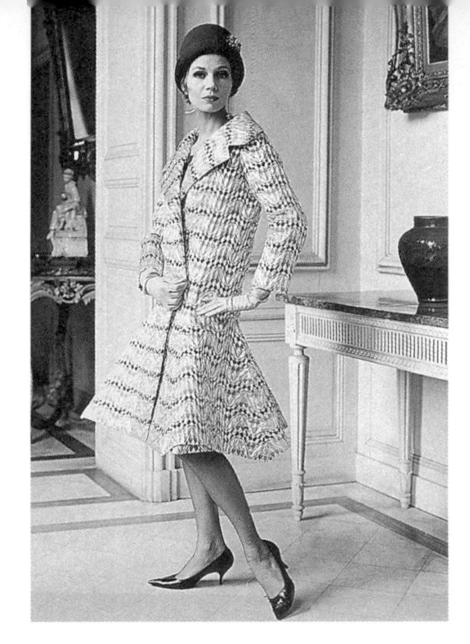

Top:
Madame Grès, ballgown in sheer nylon, 1961.
Lanvin-Castillo, dress of nylon chiffon with gigantic ruched nylon boa, 1961.
Nina Ricci, flapper evening coat in lamé (nylon with metallic thread), 1961.
Centre:
Roland Earl at Patou, an elegant suit in textured nylon and silk, 1961.
Pierre Cardin, tweedy loo< dress in acrylic, rayon and cotton mix, 1962.
Philippe Venet 'Eyelash' fringed novelty fabric in Orlon and nylon, 1962.
Bottom:
Michel Goma, lightweight coat in Orlon, 1961.
Balmain combines a coat of Orlon and Dacron with black mink cuffs, 1961.
Opposite:
Pierre Cardin, coatdress with knife pleating in Orlon with wool, 1961.

ST GERMAIN
PARIS
par
ST CLOUD

Top:
Princess coat in Orlon fleece manufactured by Millstein Inc. N.Y., 1955.
American chic seen in a stretch 'Antron' nylon dress, 1963.
Orlon suit with 'matchbox' skirt by Philippe Tournaye of Marquise, 1962.
Centre:
Anne Klein for Junior Sophisticates, cape in Orlon jersey bonded with leopard print, 1963.
Suit manufactured by Davidow in Orlon and wool tweed, 1963.
Coat designed by Jo Cope and of Pattullo in acrylic and wool, 1963.
Bottom:
Larry Aldrich design in Stern and Stern fabric.
A semi-transparent nylon net dress embroidered with gold brilliants and beads, 1963.
Ben Reig suit in 'Acele' acetate and nylon bouclé tweed by Tzaims Luksus, 1963.

Madame Grès, Maggie Rouff, Lanvin-Castillo, Nina Ricci, Emmanuel Ungaro, Philippe Venet and the fashion modernists, Cardin and Courrèges.

Du Pont was providing an astonishing – and free – fashion information service to American editors and to the public and, in the process, was giving French couturiers huge amounts of invaluable international publicity. The company was also doing its best to promote home-grown 'American Chic' by supporting and publicizing the designers attached to the New York Couture Group through advertising and gala fashion shows. In the fifties and sixties, only a few American names, such as Geoffrey Beene and Bill Blass, were able to compete with the French. Most American designers were little known in Europe, partly because the designing system and scale of manufacturing was totally different in the United States. Everything was geared to the ready-to-wear industry and great designers like Claire McCardell worked under department store labels such as Townley Frocks and for Hattie Carnegie and anonymity was the result. The era of the named designer was still to come when, during the 1980s, Claire McCardell's pre-war relaxed, casual 'American Look' was taken up by designers such as Donna Karan, Ralph Lauren and Calvin Klein.

Above: 'Espace' collection by Pierre Cardin, 1964. New materials were the last frontier in fashion.
Below: Pierre Cardin at work sculpting a surprising silhouette in fabric.

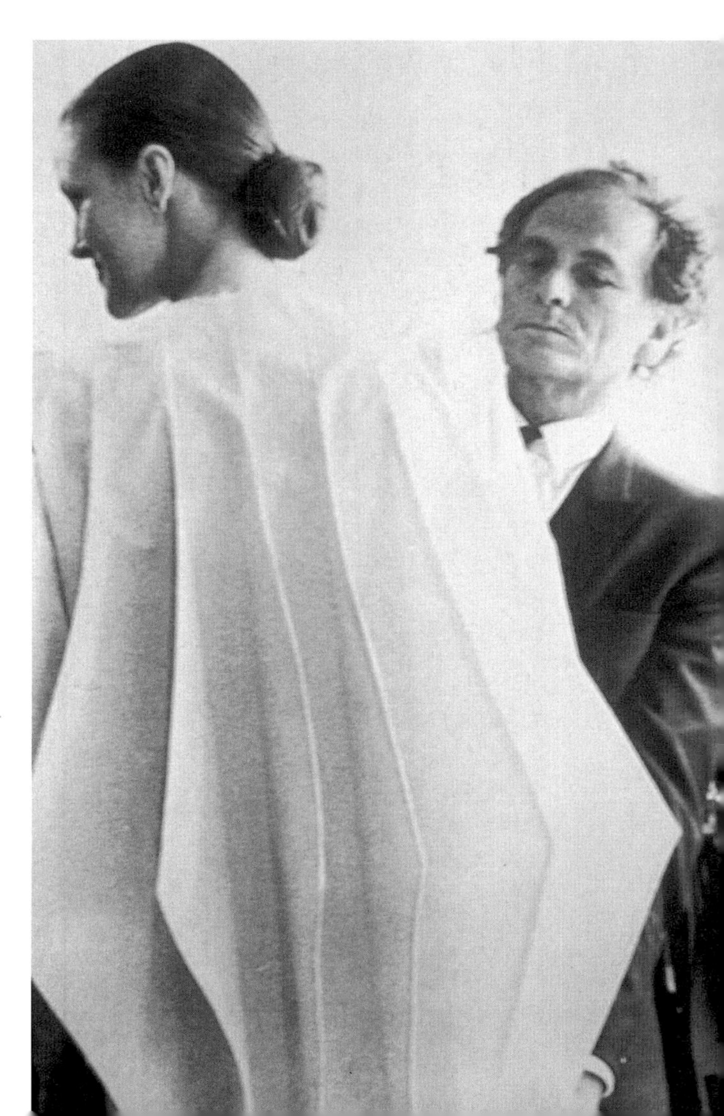

Couture Goes into Orbit

Although the fibre companies were pleased to see man-mades making inroads into traditional couture, an important international competition which would catapult synthetics and plastics into the fashion centre-stage was gaining momentum – the moon race. In 1950 the film *Destination Moon* was released, the first of many fantasies reflecting the craze for space travel, culminating in 1968 with Stanley Kubrick's classic *2001: A Space Odyssey*, written by Arthur C. Clarke. Space technology was amazing and fascinating and particularly exciting to the younger generation. Telstar, the first communications satellite, was launched in 1962, signalling the start of mass, popular culture when it brought television pictures across the Atlantic into European homes. The most potent demonstration of television's new scope came soon after with dramatic broadcasts of rocket launches and views of the earth from space, culminating with the moon landing in 1969, seen as the greatest technological challenge in history and watched live by one-fifth of the world's population in 47 countries.

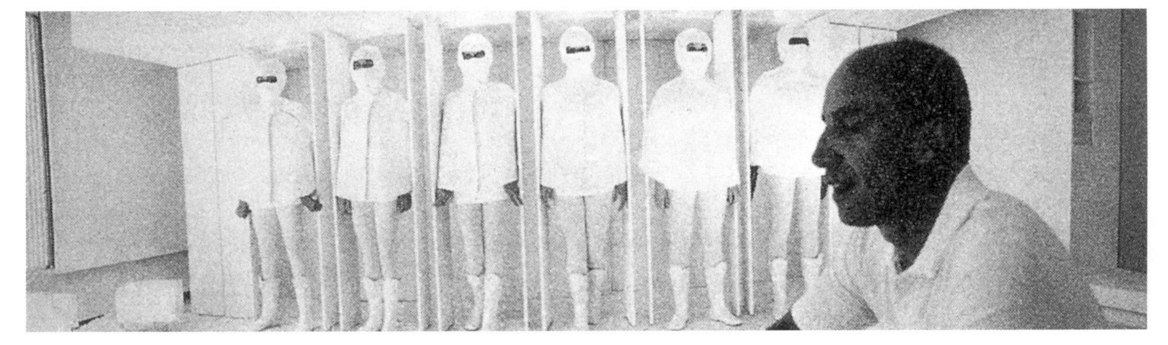

André Courrèges' silver and white 'Space Age' Collection, 1964.

(Du Pont had a hand in this triumph because 20 of the materials used in the 21-layer Apollo moon suits were originally developed by them 'for earthbound use'. These fabrics included nylon, Dacron, Lycra, Neoprene, Mylar and Teflon.)

Couture had to find a way of reinventing itself to fit into a rapidly evolving mass culture. Extravagant and socially-driven wardrobes of clothes seemed an outdated requirement and aspiration in the radicalized sixties when travel, the cinema, magazines, television and music were challenging the traditional social structure and influencing tastes. Delicate cocktail dresses, voluminous ballgowns, corsets and stiletto heels spoke conspicuously of another era and lifestyle and a more fitting wardrobe called for sharp, slim, functional, geometric shapes – suitable, perhaps, for intergalactic travel. NASA had become a household name and all things space-like held centre stage, particularly affecting fashion, the most future-obsessed of all the design professions. The space race gave fashion many possibilities, new materials, new silhouettes and, above all, a coherent new style – 'cosmic'. Couture was antiquated and rooted in the past, already dead and probably decomposing when a trio of designers reignited the Parisian torch: Courrèges, Cardin and Paco Rabanne. They adopted modern, spacey, synthetic materials and futuristic shapes. Paris managed to hold onto fashion relevance during this heady period from the moment when Courrèges launched his legendary silver and white 'Space Age' collection in 1964. World media attention was captured by his space-capsule salon and the strange robotic poses struck by his models – 'bizarre, tall things with short hair, flat-heeled boots, blinking through giant white spectacles in his laboratory-white showroom'.[13] This show represents a watershed in modern fashion.

Courrèges, for many years the chief cutter at Cristobal Balenciaga, transferred the magic of his formal tailoring skills into the creation of geometric, A-line, mini-skirted, liberating clothes, coupled with his signature white kid go-go boots. He was the leader of a group of couturiers dubbed the Space Age designers, all of whom were obsessed with the future. 'In 1965 Ungaro went on a silver binge: silver wigs; silver-soled boots; silver buttons, collars and mesh stockings. Ungaro's...clothes were always sprinkled with Space Age lunacy. Cardin, too, was considered a Space Age designer....The look was difficult to wear...yet it was photographed endlessly by the press.'[14] The press were still more excited by Pierre Cardin's 'Cosmonaut' collection of 1966, known as the 'Astronaut' look when it was being marketed in the United States. When in 1969 Neil Armstrong walked on the moon, Cardin was said to have gasped, 'You see, I was right'.[15] Cardin favoured hard edged, sculptural constructions, set off with bizarre plastic and vinyl accessories. His enthusiasm for tailoring, construction and technology in combination knew no bounds and in 1968 he even created his own bonded fabric, 'Cardine', which resembled moulded egg-boxes. He probably devised this fabric in collaboration with Union Carbide and some idea of the relationship between designer and corporation was given in a report to Du Pont: 'Cardin worked with Carbide against a flat fee "rumored" to be $20,000 – plus free fabric, etc, and has shown only two of their garments in the show the writer saw, although he is said to have cut 10 of them. Cardin's collection was, as usual, the largest of all couturiers (240 garments) and judged by the press and trade a very good one.'[16]

The best way to belong to the future was to look as if you were dressed for it, by wearing the weirdest futuristic clothes available. Designers adopted a

Top: Cardin's outfits for would-be moon maidens and cosmonauts.
Above, left: Bonded synthetic fabrics and PVC ensure the wearer wrinkle free space travel.
Above: The much-copied go-go boots designed by Courrèges.

Above: Emmanuel Ungaro's A-line coats with geometric trompe l'oeil effects in vividly dyed Orlon, 1966.
Below: Paco Rabanne in 1966 with two of his 'engineered' evening dresses made from linked discs of phosphorescent plastic. Pliars and wire were the new tools of the fashion designer.
Right: Courrèges turned everyday plastics into sixties haute couture. Details show a sleeve from early fibre optics, a boldly plastic zip, a gown made from telephone wire and a dress made from melted bin liners.
Opposite: Op art fabric and Space Age fashion come together in 1966 in *Nova* magazine.

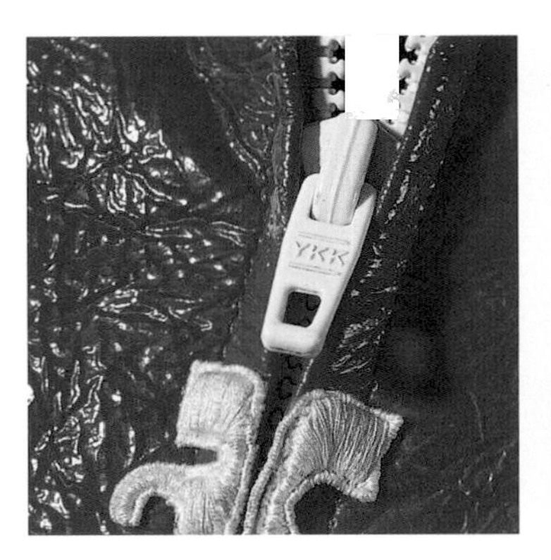

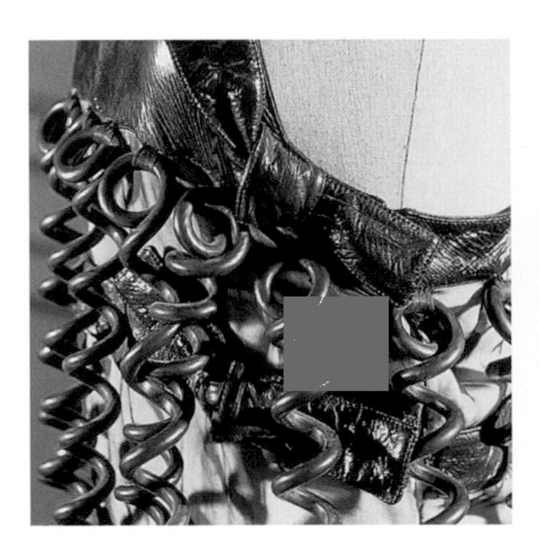

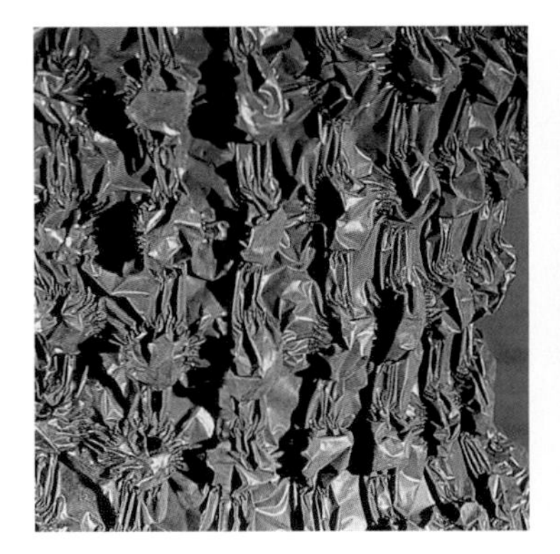

ERNSEHEN

completely new range of dress materials, all forms of synthetics and plastics: silver-finished fabrics, resembling the tin foil developed for insulation in outer space; bonded nylon jerseys, as rigid and geometric as a landing module; polyvinyl chloride (PVC); and plasticized aluminium (Lurex). Fashion pages of magazines and newspapers illustrated this space fixation throughout the sixties with countless examples of silvered-plastic coats, white plastic go-go boots, Lurex stockings and moulded-foam astronaut-helmet hats. 'Sewing is an antiquated bondage, in future all garments will be melted or welded together', declared Paco Rabanne, another great sixties futurist. Filled with optimistic visions of the wonders that technology could provide, he not only embraced new materials but fabricated futuristic garments by devising a system of linked plastic discs. Abandoning tradition, Rabanne believed that the only way fashion could progress was by discovering and using new and alien materials. Preferring to define himself as an engineer, he became famous for his mini dresses, constructed from unlikely materials such as colourful Rhodoid plastic, metals or aluminium.

The Texan designer, Ruben Torres, was perhaps the most future-obsessed and least documented of all the Space-Age designers then working in Paris. His utopian concept was of a near future when technology would render the hand-made products of couture obsolete by providing equally high standard garments for the masses. He designed 'clothes for the man of tomorrow – to be worn today'. His futureman would be dressed in a functional stretch jumpsuit, whether he was a 'new-look labourer' or 'tomorrow's city man'. Listing the factors that were influencing the clothing revolution, he included developments in textiles and manufacturing; 'man's future lies in his own inventions and less and less in those of nature'.[17]

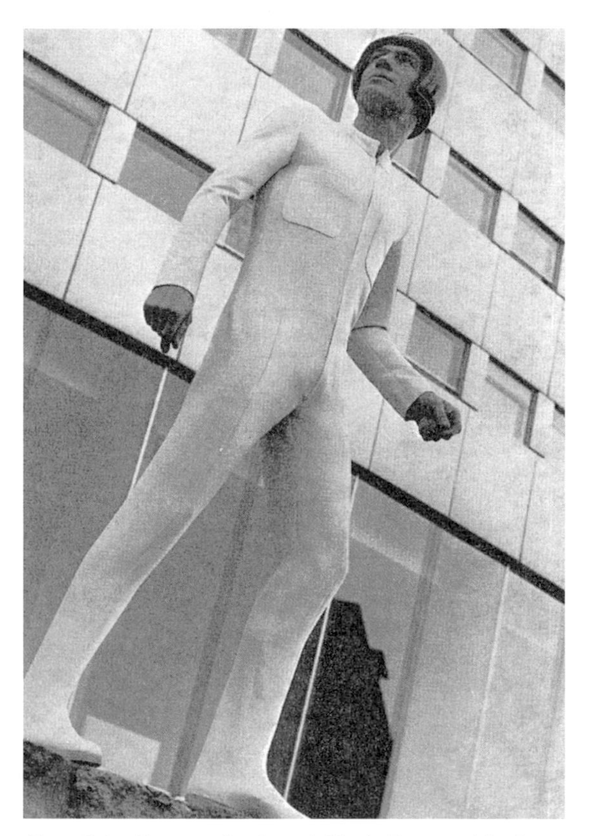

Above: Ruben Torres predicts future fashion in 'Tomorrow's Look Today' – a one-piece space-inspired stretch jumpsuit of 1967. Opposite: Dress in Cardin's patented moulded fabric 'Cardine', worn with PVC accessories in 1968.

Hopeful Futuristic Fibres

Across the Atlantic, Du Pont was busy with the launch of several newly developed nylons including 'Sparkling' and 'Cantrece', both hosiery fibres produced just in time for the leggiest decade of the century. As is often the case when designers are very much in tune with their times, similar silhouettes and ideas were emerging simultaneously, and the launch of the mini skirt has been credited to Courrèges, Mary Quant and Pierre Cardin. Whoever introduced them, mini skirts were the fashion essential and tights, in infinite colours and patterns, made great fortunes for all the synthetic-fibre manufacturers. 'Qiana' was another new nylon brand destined for the fashion industry and one which Du Pont tried to introduce through the Parisian couture industry. A snapshot of the commercial tensions and marketing mechanics of launching a newly branded nylon is preserved in a report made for Du Pont in 1968 which tracks the success of Qiana with the major Paris couturiers. Du Pont clearly expected to buy its way into couture with its latest new nylon and herein lies a tale of bargains struck and deals turned down.

First, the names of the fabric houses weaving the Qiana fibre – Abraham, Bianchini, Bucol, Nattier, Staron – were set against the precise number of 'major' and 'minor' designers using their textiles. Du Pont had already sponsored a Qiana wedding gown designed by Marc Bohan at Dior, valued at $4,437 and given to the Metropolitan Museum of Art in New York. Bohan succeeded Yves Saint Laurent as chief

Above: Du Pont launch yet another new nylon, 'Cantrece', in 1964, destined for the hosiery department. Silk stockings were already a dim memory.
Below: Picture opportunities, glamour and sales went hand in hand. With smiles all round, 'Miss Cantrece Nylon' is measured for a pair of new wrinkle-free stockings.

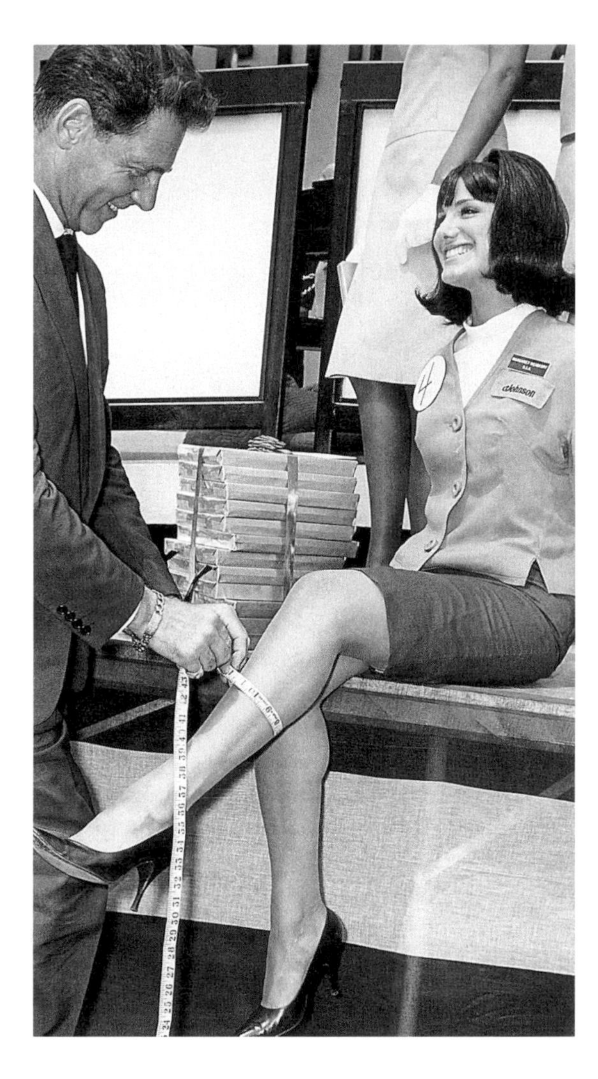

designer and Artistic Director of Dior in 1960 and Saint Laurent established his own label in 1962. Maison Dior insisted that Du Pont assist them financially with two shows in the United States – the cost of a New York show at the time was somewhere between $15,000 and $20,000. It is clear from the report that Dior's directors believed that a financial deal had already been struck between Du Pont and Saint Laurent.[18] It was the writer's opinion that unless satisfaction was given to Dior 'we will be back to the old status of bargaining for each garment, photograph, advertisement, entrance ticket, etc'. 'Ungaro...made a major effort in Du Pont fibres, 10 Qianas and several "Dacron"/worsteds, [but he] absolutely wants us to finance a ten-minute film on "How he works, how he selects and conceives fabrics, and how he shows them on his premises as garments"....Unfortunately, [this] may cost as much as $20,000 to $30,000. In spite of this, acceptance of Ungaro's proposal is strongly recommended...It is understood that we could have any desired number of garments from Du Pont fibres in such a film.'[19]

Qiana was already causing problems: to avoid melting or scorching the fabric, iron temperatures had to be kept very low, and severe static was reported by Balmain's salon where a wedding dress in Qiana was said to have 'electrocuted' the model.[20] In addition, rebellion was abroad among some designers; Patou accused Du Pont of wanting to use the couture industry as a Trojan Horse to introduce their new fibre in a less expensive way, through editorials instead of through endorsed advertising.

As the fibre wars accelerated in the early sixties, the Kings of Couture could command considerable returns for their fibre endorsements and some couturiers demanded larger fees than Du Pont was prepared to pay. The pages of French *Vogue* or the British *Nova* amply illustrated the fibre companies' courtship of designers – they were filled with a bewildering range of both synthetic and natural fibres, sometimes being promoted by the same designer.

40 Million Home Couturiers – and the Slide Begins

The ultimate democratization of couture was an American aspiration and this would only be possible, Du Pont believed, with the help of new synthetic fabrics that simulated glossy satins, fine silks and luxurious wools. A fashion-conscious mass-market

Emmanuel Ungaro gives 'Pure New Wool' a modern couture image during the mid-sixties fibre wars. French *Vogue*.

Robe de CHLOÉ en gaze lunaire de J. B. MARTIN
TERGAL

Polyester claims its own trophy designer, Chloe, working in 'Tergal', the brand name of Rhodiacata. French *Vogue*, mid-sixties.

was a boon to the synthetic-fibre industry where the key players, Du Pont, ICI and Courtaulds, needed to sell increasingly large volumes of their raw fibres. One significant way of expanding the taste for synthetic fabrics even beyond high street ready-to-wear was to appeal to the home dressmaker. Du Pont in particular made efforts to reach out to this enormous market; an article in the *Du Pont Magazine* in 1966 was entitled '40 Million Home Couturiers', the number of American women who had 'an intimate knowledge of the sewing machine'. The article stated: 'Surveys reveal that every eighth dress worn today is a do-it-yourself creation. Moreover, its makers are the mainstays of an industry that last year grossed $1.35 billion – up significantly from the $1 billion spent in 1960. A rough breakdown of outlays in the 1965 home-sewing market: fabrics, $750 million; accessories, $300 million; sewing machines, $180 million; patterns, $70 million; and thread, pins, needles, etc, $50 million.'[21] Such statistics were irresistible to the fibre-marketing men, especially when they were being told that 'the surface has barely been scratched'. There is no doubt that the home sewing market at the time was vast and companies such as Cohama Fabrics eagerly targeted home sewers in their 'Thimble Couture' programme, designed to sell their Du Pont stain-resistant fabrics. Remembering that this article was written when the feminist movement was beginning to take hold in America, its declared intention was to 'trigger the instinct for garment design latent in most women, be they budget-bound secretaries or pillars of the smart and moneyed set...the yen for do-it-yourself chic knows no artificial boundaries'.[22]

Even the wife of the Vice-President of the United States, Mrs Hubert Humphrey, made the dress she wore at his inauguration; 'when a woman learns to sew, she becomes more fashion-conscious than if she just goes out and buys what she wants', Mrs. Humphrey told a *Women's Wear Daily* reporter.[23] The same piece stated that sales of McCall's patterns had been increasing at a rate of 10 per cent a year since the beginning of the 1960s thanks to the involvement of designers such as Pauline Trigère, Geoffrey Beene

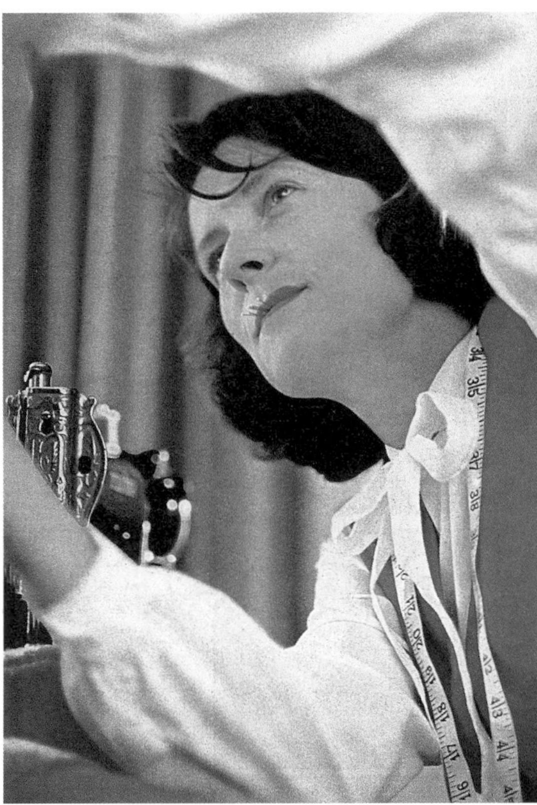

and Bill Blass. A Pauline Trigère evening dress illustrated in the article would have sold for $300 but could be home-made in polyester for a material cost of about $16. The do-it-yourself couturiers continued to be courted by Du Pont well into the 1970s and in 1972 home sewers were targeted with Qiana. Herein lies the symbol of synthetics' social decline: from the Dior nylon gowns of the mid-fifties to the Qiana home-mades of the seventies.

Somewhere in the moon's vast wilderness is planted the flag of the United States of America – the ultimate symbol of national triumph and scientific progress – eternally suspended in zero gravity. And, appropriately for the era, the flag is made from indestructible, standard issue Du Pont nylon. Mankind's 'giant leap' was the greatest visible demonstration yet seen of the power of technology. American space technology and fibre science brought about a total transformation in Parisian couture. Synthetic materials inspired its new shapes and silhouettes and for the first time the dictators of fashion were responding to a new form of popular, youth led culture. But, ultimately, couture was destabilized in the egalitarian pop era and its influence was reduced to that of a few designers, mainly the technology enthusiasts, Courrèges, Cardin, Ungaro, Rabanne and Torres. 'Space style', artificial materials and new techniques such as vacuum-forming and heat-moulding gave couture a veneer of modernity – just enough to keep the press interested. It was a short-lived and rare marriage between haute couture, mass culture and synthetic fabrics. It could be argued that the creations of the Parisian Space-Age designers were not clothes but costumes for a fantasy futuristic life, but couture design is largely about the making of a showpiece. Despite the similarity of the basic tunic-form, fabrics and the A-line silhouettes between Paris couture and London's radical fashions, the two spheres were worlds apart in intention. For British designers, synthetics were a modernizing tool – the fabrics that could make the cheap, short-lived, fun fashions of street style.

Opposite: Edwin 'Buzz' Aldrin uses a Du Pont nylon flag to claim the moon for America.
Left, top: Du Pont compared the price of a designer original in silk with a line for line copy in polyester. This evening dress and stole would sell for $300 but could be home made for $16-worth of fabric.
Left, bottom: One of America's estimated 40 million home couturiers, 1966.

CPL 91

5

> 'The dress designer of today who wants to last until tomorrow needs to be aware of all the untapped beauty lying yet unexplored in plastics and in laminated and bonded fibres – in fact, the time is ready for the new men or women truly of the moment. . . . Technically, the dress designers of the future will probably need a basic knowledge of science and ergonomics and will almost certainly mould their garments, both in natural and man-made fibres, rather than cut and sew them.'
>
> Janey Ironside, Professor of Fashion, Royal College of Art, *Queen*, 21 June 1967

1955–70

The Fabric of Pop

In May 1966, *Time* magazine coined the phrase 'Swinging London'. As the world capital of pop and fashion, the city was at its peak. The model, Twiggy, was voted Woman of the Year and Mary Quant had worn a mini skirt to Buckingham Palace when she collected an OBE. London was fermenting its very own fashion and rock revolution and for the first time the word 'culture' was prefixed with 'mass', 'popular' and 'youth'. 'I was born with a plastic spoon in my mouth': The Who's hit record *Substitute* captured the spirit of the times. Substitution was everywhere; 'pop' had replaced elitist art, plastics were the materials of the moment, and Britain's cherished class system was inverted when, for the first time, 'working class' attitudes and accents acquired a social and fashion cachet. In the 20 years since the war, Britain had changed beyond all recognition.
The tweedy conservatism and austerity of the post-war years seemed like a dim newsreel compared to the technicolour lifestyles created by the baby-boom generation, a culture made possible and carried forward by synthetics and plastics; from vinyl discs to Courtelle knits and inflatable furniture, chemical materials literally shaped 'pop' style and fashion.

Plastic Power and Youth Culture

Traditional materials like wood, stone, glass, ceramics, metal and even wool imposed limitations of form on the designer but now all of these substances could be imitated or substituted with plastics –

Opposite: PVC scooter suit, designed in the Fashion School at London's Royal College of Art in 1965.

Life before the youthquake struck Britain. Roger Mayne's photograph *Mother and Daughter shopping*, 1956.

plastics could be made to do just about anything. 'The hierarchy of substances is abolished', wrote Roland Barthes in 1954, instead, 'a single one replaces them all: the whole world can be plasticized'.[1] Barthes anticipated both the coming ubiquity of plastic objects and marvelled at the infinite number of possible transmutations of a single substance. 'Despite having the names of Greek shepherds (Polystyrene, Polyvinyl, Polyethylene), plastic...is in essence the stuff of alchemy. Its quick-change artistry is absolute: it can become buckets as well as jewels.'[2] This 'miraculous' substance was both a 'wonder' and an 'enigma'; with neither a past nor a tangible identity, plastics were neutral in value. This was their advantage – and their downfall – because their worth was entirely socially constructed. It is the parallel story of plastic's cousins, synthetic fabrics.

A passion for clothes was unleashed in the wake of wartime rationing and all the gloomy government slogans instructing the public to 'Make War on Moths', 'Choose All-the-Year-Round Clothes' and 'Ask Yourself, Which Clothes Can You Do Without?'. If ever there was an invitation to excess this was it. The young in particular longed for change and variety and this urge reached its apex with the throwaway fashions of the 1960s. Change was everywhere in post-war Britain, new housing, the relocation of communities, a new leisure industry, television, teen magazines, holidays abroad, launderettes, hire purchase and the first supermarkets, all of which reflected the beginnings of an affluent society. London was the centre of the first youth culture, a newly invented synthetic lifestyle described by Sylvia Katz in *Classic Plastics*: 'Perhaps the inhabitants of the espresso bars of the late fifties and early sixties were really the first true plastics generation, with their Terylene shirts, moulded (pleated) polyester skirts, beehive hairdos sealed in position with vinyl acetate lacquer and legs encased in "Sheerer Era" nylon stockings suspended from synthetic rubber roll-ons.'[3]

Loot to Spend at Last

The odd behaviour of a previously unknown group, 'teenagers', had first begun to irritate and then to challenge the British Establishment in the early 1950s. A rebellious and alien culture was materializing and generational hostilities soon came to the surface. Disturbing new words entered the everyday vocabulary – 'juvenile delinquent', 'rock and roll', 'fans'; youth became a 'problem' phase of life and group identities became centred around musical styles. The teens in former generations had been without voting rights, without disposable incomes and without music or an identifiable culture of its own. The photographer Roger Mayne captured the claustrophobia of his times in a 1956 documentary photograph, *Mother and Daughter Shopping*.[4] There is little or no distinction between the style of clothes worn by the pair and this was largely true of all social classes in the early fifties; debutantes and their mothers would have clothes made at Lachasse or Norman Hartnell while working girls and their mothers would buy ready-to-wear from department stores or rely on the skills of their local dressmaker. In terms of style and design, the generations were as one. The status quo was challenged in the mid-fifties and overthrown in the sixties.

'Street style' is synonymous with British fashion and it also had its roots in the fifties. Immediately after the war, housing was an acute problem. This was a time when most young people lived at home, often for years after their marriage. There was little alternative but to meet and socialize on the streets and Teddy Boys were disparagingly known as 'Corner Boys' as they spent much of their time on street corners. The street was their meeting place and their social centre, the eternal showplace for youth styles. *Absolute Beginners*, the novel by Colin McInnes published in 1959, marks the transition between the old life of the submerged young and the new world of the emancipated teenager. He writes of the 'savage splendour' of the days when teenagers 'had loot to spend at last'.

In the days of rationing, consumers did not have the luxury of choice, but by the mid-fifties the manufacturer had to be fairly confident about what the consumer wanted before he committed himself to making products, particularly yarn producers such as ICI, Courtaulds and Du Pont as the chain from their product to shop counter took at least 18 months to complete. Consequently, fibre manufacturers were trailblazers in market research, keen to know and understand the fashion inclinations of consumers. While Du Pont published the *Du Pont Magazine* and *Better Living* to disseminate information about its products and consumer spending habits, ICI produced a series of publications during the fifties and sixties promoting British synthetics: *ICI Fibres Magazine*, *Nylon Magazine* and, through British Nylon Spinners, *BNS Magazine*. These included the findings of market research and threw light on key questions: was any notice taken of magazine advertisements; were opinions shaped by friends, shop displays or television; were shirts bought by men or their wives; were laundering considerations or price the main issue; and did consumers know – or care – about the differences between nylon and cotton? The main article in a 1958 issue of *BNS Magazine* was called 'Consumers Pay the Piper – What Makes them Like the Tunes?'. The task given to ICI's researchers was to uncover social attitudes to nylon's personality. Consumers were shown photographs of typical houses and were asked: 'Do you think a woman living here would buy a nylon blouse?'. In the results of the survey, nylon was found to be well up in the social hierarchy of fabrics.

Mark Abrams' *Teenage Consumer Spending in 1959* was the first of many statistical and analytical surveys to record the habits of consumption of the newly classified spenders, teenagers.[5] A narcissistic obsession with clothing was one way for urban, working-class teenagers to assert themselves. Fashion became a

"They might not be able to afford nylon"
2 "She's young and modern—she'd buy nylon"

"She looks smart—she'd like nylon and could easily afford it, too"
4 "I think she's middle-aged. She must be well off; but she'd think nylon is for younger people"

Above: A British Nylon Spinners survey set out to define which social categories of women would buy a nylon blouse. The results were published in *BNS Magazine* in 1958.
Below: Two ultra-modern girls photographed in 1963 at an ICI factory to celebrate the fashionable wonders of nylon yarn.

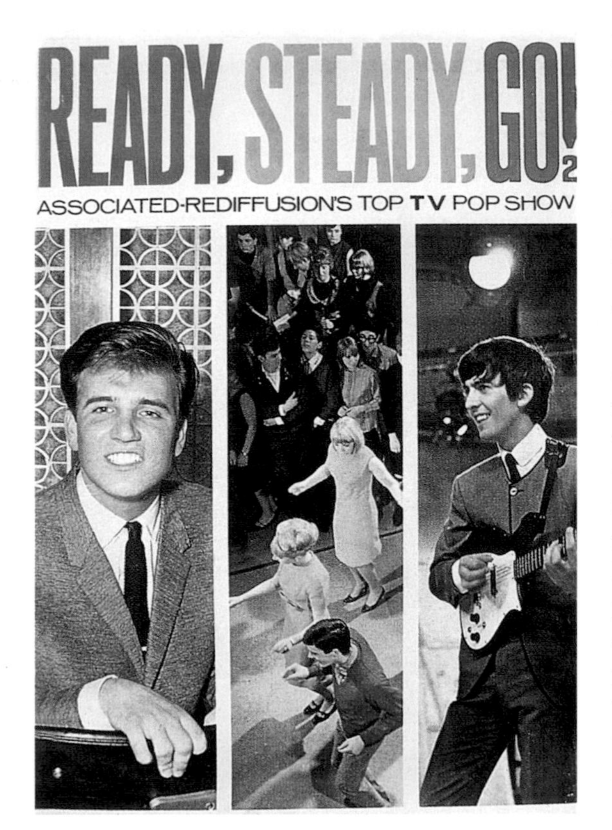

Pop culture gets under way with its own magazines, television programmes, pirate-radio stations and independent fashions.

vehicle for teenage social identity, with any number of styles signalling a set of behaviour patterns and cultural affinities particular to each one. Abrams calculated that the real spending power of teenagers had almost doubled between 1938 and 1958 and their disposable income continued to increase well into the next decade. Teenagers became consumers of the entertainment and fashion culture they had created and Abrams made a breakdown of their expenditure: 'a good share goes on "pop records", gramophones, romantic magazines, fiction paperbacks and cinema, but the quite large amount of money at the disposal of British teenagers is spent mainly on dressing up in order to impress other teenagers and on goods which form the nexus of teenage gregariousness outside the home'.[6]

The 'teen' phenomenon was destabilizing markets in the United States as well as in Britain. In 1965, the *Du Pont Magazine* covered the social change in an article called 'Teen Age Boom Rocks the Market', reporting that: 'Never before has U.S. business been more preoccupied with the indescribable and ubiquitous phenomenon known as the teenager. The nation's population between ages 13 and 19 has mushroomed since World War II to a mighty 24 million and, although they're barely out of childhood, these youngsters are, in the main, sophisticated, informed and discriminating consumers. As such, they are strenuously wooed by businessmen who recognize the awesome power they and their whims can exert in the marketplace.'[7] In 1964, American teenagers spent $5 billion on clothing, 'and to an astonishing degree, girls buy independently of their parents...selecting what takes their eye or what looks glamorous or what they think their peers are wearing. Virtually all apparel manufacturers agree that the teenager is one of the most important influences in the fashion world today.'[8] But, although Du Pont was well aware of their potential new market, they were out of touch with the anarchical ethos that made Britain such a forceful pacesetter. In 1963, Du Pont, in collaboration with the New York fashion designer John Weitz, came up with the distinctly tame concept of the 'Well-Mannered Look' for teenagers.[9]

Pop music was the glue that held youth culture together and fashion defined the various 'tribes' within it, but the whole creative phenomenon depended on a set of circumstances that young British people were enjoying. First, there was the newly affluent young consumer, seeking an identity that would signal their particular sub-cultural values; second, there was for the first time a pool of art-school trained young designers who were eager to break away from tradition by using the newest materials; and finally, there was the 'boutique', the direct link between the designer and the consumer. Everything was in place to bring about the fashion revolution made by and for the young. Clothes were now ephemeral and relatively cheap and synthetic fabrics and dyes brought in a range of lurid, artificial, nightclub-neon colours that could be bold and shocking for the few weeks the garment might be worn. Luminous lime green, sharp yellow and hot pink lent adventure and youthful excitement to clothes. These were the vivid colours of Rock and Roll and the unease they instilled in adults was exacerbated by a commonly held belief that luminous fabrics were radioactive, which led teachers and parents to confiscate and destroy these 'dangerous' items, often socks.

Boutique Fashions from the Bedsit

Perhaps the greatest boost to London design was the coming of the boutique, the small independent retail outlet that meant that designers could also be their own retailers. For once, clothes designers had the advantage: by being in control of both production and the point of sale they had liberated themselves from the need to compromise their work to satisfy conventional in-store buyers. This freedom of expression allowed them to experiment in an unprecedented way with new materials and designs. A few wild ideas could be stitched together at home and tried out in the local boutique. A profusion of fashion boutiques – Quorum, Foale & Tuffin, John Stephen, Mr Fish, Mr Freedom, Countdown, Palisades, Annacat and many others – were soon clustered around the King's Road and Carnaby Street, a colourful part of the new retailing culture in London that had a very strong design emphasis – Habitat was the huge and immediate success story in home furnishings.

Mary Quant's Bazaar, the first boutique, had opened on the King's Road in 1955, jointly financed by Alexander Plunket Greene and Archie McNair who was the owner of Britain's first coffee bar, also on the King's Road.[10] Quant was the pioneer of fashionable clothing for the younger generation. Her philosophy was quoted in the *New York Times* magazine: 'the young should look like the young. . . . The old could, if they wished, look like the young, but the young must not on any account look like the old.'[11]

Traditional British couture studiously avoided synthetics because its clients held firm to the prejudice that they were inferior and low-class fabrics, a view that still makes many of the current generation of couture clients recoil in horror at the mention of nylon or polyester. Synthetics just did not sit comfortably with the British couture ethic, but it was a totally different story at the younger end of the market where they were the most relevant design fabrics. Never had the generations been so polarized. In 1967, Janey Ironside, Professor of Fashion at the Royal College of Art in London, wrote: 'If "all that the young can do for the old is to shock them and keep them up to date", as Bernard Shaw once said, they have been fulfilling that function with enthusiasm recently, and fashion in clothes has been out there in front getting its share of the blame as usual. There is something about girls in mini skirts and boys with

Meet today's teen. She's the girl with the well-mannered look. Very big on classics, clubs and careers. Her clothes? Very in, very "Orlon"!
Carol Rodgers tailors the teen-size classic in a blend of "Orlon" and rayon. Soft, this blend is. And luxurious. And it's for you in a kick-pleated twill, in grey or red. Or a bold bash of a plaid, in blue or brown. Sizes 5-15. Each, about $12. Both dresses at Davison's, Atlanta; L. S. Ayres & Co., Indianapolis; The Broadway Valley, Los Angeles; Macy's, New York; Woodward & Lothrop, Washington, D. C.; and other fine stores throughout the country.
DU PONT

Above: Du Pont invited the neat, sweet 'teens' of America to adopt a 'Well-Mannered Look', created in acrylic and polyester.
Below: A new synthetic lifestyle for the teenager in a 'sun-ray' permanently pleated polyester skirt, surrounced by vinyl records, played on a plastic record player.

Above: Marion Foale and Sally Tuffin with staff at their boutique in Carnaby Street. Indepedent retail was the key to success for London's young designers. Photographed by the designer James Wedge.
Right: Boutiques opened up all over Britain. By 1966, plastic versions of Courrèges' go-go boots were everywhere, including these seen in trendy Halifax.

long hair that releases an instant animosity from older people who could not possibly wear them.'[12]

After graduating in illustration from Goldsmith's College of Art, Mary Quant worked with a London couture milliner, Erik, spending 'three days stitching a hat for one woman'. During this time it struck her that fashion should not exist for the privileged few but for everyone and especially for the young. With no formal fashion training, she began to make clothes in her bedsit. 'I had always wanted young people to have a fashion of their own', her autobiography records, 'absolutely Twentieth Century fashion...chosen by people of their own age....The young were tired of wearing essentially the same as their mothers.'[13] Fashion had not changed for decades, it was dictated by Parisian couturiers and only 'pallid and poorly made couture derivations were available to each descending economic echelon'.[14] By 1961, another Bazaar had opened in Knightsbridge and Quant went wholesale in an attempt to keep her prices within reach of the mass market. Conscious of the growing spending power of subcultural groups, she saw that it was the Mods who were the driving force that uprooted the stagnant dress trade because 'there was always a need to keep pace with them....The voices and culture of this generation are as different from those of the past as tea and wine and the clothes they choose evoke their lives...daring, gay, never dull'.[15] Mods (an abbreviation of 'moderns') epitomized the 'London Look': girls in Mary Quant's PVC and bonded jersey tunics with thigh level mini skirts and boys in Italian-cut polyester suits. The era of the coded fashion uniform had arrived. Spending their weekends shopping in boutiques, riding around on Lambrettas and listening to The Who, theirs was the lifestyle of the teenage consumer personified.

The arrival of the young peacock male fuelled the sixties fashion boom. Out went safe suitings and in came florals, frills and novelty fabrics.

Reinventing Menswear

Fashion became a fast, capricious business in the sixties, stock changing almost weekly in the liberating environment of the designer-owned boutique. Menswear underwent a complete transformation; the Burton suit was no longer enough, instead the times called for the peacock male. John Stephen, known as the Million Pound Mod and the King of Carnaby Street, helped to remake the male fashion scene. The dullness of fifties menswear was made all too real to Stephen when he was working in the men's department at the Glasgow Co-op. In 1957 he opened the first of the fashion shops in Carnaby Street; before that 'there were no retail businesses [there] except a tobacconist'. It was an ideal location, 'cheap rents, little traffic and close to Regent Street and Piccadilly'.[16] 'I hate Paris', he said in the mid-sixties, 'The fashion people there have been upset by Carnaby Street. The couturiers worked together to make Paris a fashion centre. There was little creative thinking or individuality in their merchandise.'

In the eyes of the world, London itself became Carnaby Street and the King's Road and boutiques sprang up by the dozen. Before long, John Stephen was counting The Beatles and the Rolling Stones among his customers, plus many of their fans who all wanted to emulate their look. The Kinks hit song, *A Dedicated Follower of Fashion*, mocked the fashion fever that had suddenly afflicted the male population. All the gender dressing taboos were pushed aside. John Stephen recalled: 'When I started I had to fight to sell our clothes. People laughed at pink and red slacks. They said they were clothes for women and were

Above: Mary Quant's PVC 'Wet Look' collection, 1964.
Opposite: Twiggy with painted vinyl dress by Joan 'Tiger' Morse, 1967.

effeminate. Nobody should wear them. We have broken through the barrier and have made the public realise that there is such a thing as fashion for men and that they should be able to wear what they like. There is no longer a class thing any more about clothes. In the pub or the dance hall you can't tell what strata of society people belong to by the way they dress. It gives me very great pleasure to visit different countries throughout the world and know they are all looking to Britain for a fashion lead.'[17]

Many new young designers like John Bates and Jim O'Connor blended plastics with synthetics and borrowed images from pop iconography. Menswear literally blossomed, into floral and paper shirts, bright matching ties and belts, PVC coats and one-piece zip-up suits. John Weitz, a former racing driver and known as America's Pierre Cardin or Hardy Amies, was also placing his fashion bet on the throwaway solution: 'Shirts will inevitably be disposable, of paper; also underwear, socks and even swimwear as well. I foresee supermarkets for clothes...It won't happen for five or six years and it will need a new marketing set-up with a depth of production that enables garments to be sold for a few pence or shillings. Self-service vending machines will probably be the answer. It's no use getting sentimental about clothes. People want space in their homes, not storage space. In future men will buy clothes by the week. It will be good for family life and men will love their homes because there'll be much more space around. Wardrobes will have to go. And suitcases are another anachronism.'[18] Written some thirty years ago, this vision of a time when clothes would be like Kleenex seems far off the mark now, but at the time alternative, highly disposable materials seemed to herald the end of the old fashion system.

PVC Dreams and Disasters

PVC[19] became *the* signature fabric for Mary Quant, but this fashion success story was not without its teething problems. In the early 1960s Quant launched 'The Wet Collection' in polyvinyl chloride, a material invented in 1884 and chemically related to linoleum, which had never before been used in fashion.
Mary Quant described the agonies that accompanied innovation which gives a good insight into the hit-and-miss nature of sixties fashion experimentation:

'In polyvinyl chloride we had used a revolutionary new material before anyone had had time to find out and solve the difficulties of mass production. We were not the first to find out that it doesn't always pay to be first in the field.... We began to come up against the most ghastly difficulties. All the engineering work that ought to have been done before the clothes were shown...the experiments on welding seams...had been overlooked. When the stuff was put on an ordinary machine, the vinyl stuck to the foot or melted, and when we found a way of stopping this, the seams were perforated like those of a postage stamp and ripped at the slightest provocation. It was obvious that welding was the only possible process of manufacture and there was no suitable machinery for this. We ought to have been in touch with one of the big firms experienced in making macintoshes but we were still too amateur in our approach to have looked this far ahead.

We had so much publicity that the clothes were in enormous demand and we simply could not deliver. We fell down all along the line on production. It was disastrous. Everyone wanted our things and we did not know how to make them....It took us nearly two

PVC, formerly used as oilcloth and similar to linoleum, became one of fashion's favourite fabrics for clothes and accessories.

years to perfect the process whereby the seams of PVC can be held fast in a satisfactory way and by that time other designers on both sides of the Channel were as bewitched as I still am with this super shiny man-made stuff and its shrieking colours, vivid cobalt, scarlet and yellow, its gleaming licorice black, white and ginger.'[20]

Within two years of starting in business, Mary Quant had signed a design contract with J. C. Penney in New York, had formed the Ginger Group and had gone into mass production with her look. Her success was such that she was given the *Sunday Times* International Award for Jolting England Out of its Conventional Attitude Towards Clothes. By 1963, Mary Quant was designing, wholesaling and manufacturing fashions for the teens and twenties and hers was the Dress of the Year selected by the Museum of Costume in Bath. She also introduced an enormously successful cosmetics line. For all these endeavours she was voted Woman of the Year and won the prestigious *Sunday Times* International Fashion Award, an honour previously reserved for couturiers. Famous fashion designers were – and still are – usually mature males and there was something startlingly fresh about a young woman designing clothes for her own age group and for working girls' lifestyles. In 1965, she took 30 outfits and her own British models, all with Vidal Sassoon geometric haircuts, on a whistle-stop 12-city tour of the United States. The 'London Look' took America by storm.

A Perfect World in Plastic

Stuart Ewen's study of contemporary society describes the sixties as the point from which an 'alternative' or 'oppositional' culture materialized, concerning itself with 'questions of war, the environment, racial and sexual equality, global inequities, and of an overtly commercialised and superficial consumer culture'. Along came the 'rejection of the prevalent values and iconography of the primarily white, "middle-class" consumer culture…[replaced by] renegade styles, verbal expressions, ways of dress, music, and graphics'.[21] Both art and fashion found their inspiration in the new supermarkets, the fast turnover of stock, bold graphic branding and the materials of packaging – plastics and paper. The fashion revolution stood shoulder to shoulder with Pop art, the movement that began in fifties' London at the ICA (Institute of Contemporary Arts) and that was propelled into the 1960s by the writers Lawrence Alloway and Reyner Banham and the artists Richard Hamilton and Eduardo Paolozzi. Richard Hamilton's famous 1956 collage, *Just what is it that makes today's homes so different, so appealing?*, incorporated all the elements of transient, popular culture (products, the media, food, space travel, logos, pulp romance and so on) and elevated the banal to the meaningful. Pop unified everything: 'commercial' art became fine art and fine art translated itself into textiles. Acrylic paint was *the* medium of pop artists, instantly dry, water soluble, vividly coloured and conveniently packed in plastic tubes; it was also many times cheaper than traditional oil paints. The chemistry of acrylic paint was more or less the same as that of an acrylic sweater or an acrylic coffee cup and the consumers of all three were the teenagers and young adults of the baby boom generation. During the 1960s, knitwear was at its fashion peak and the competition between the producers of Orlon, Acrilan and Courtelle acrylic fibres was intense.

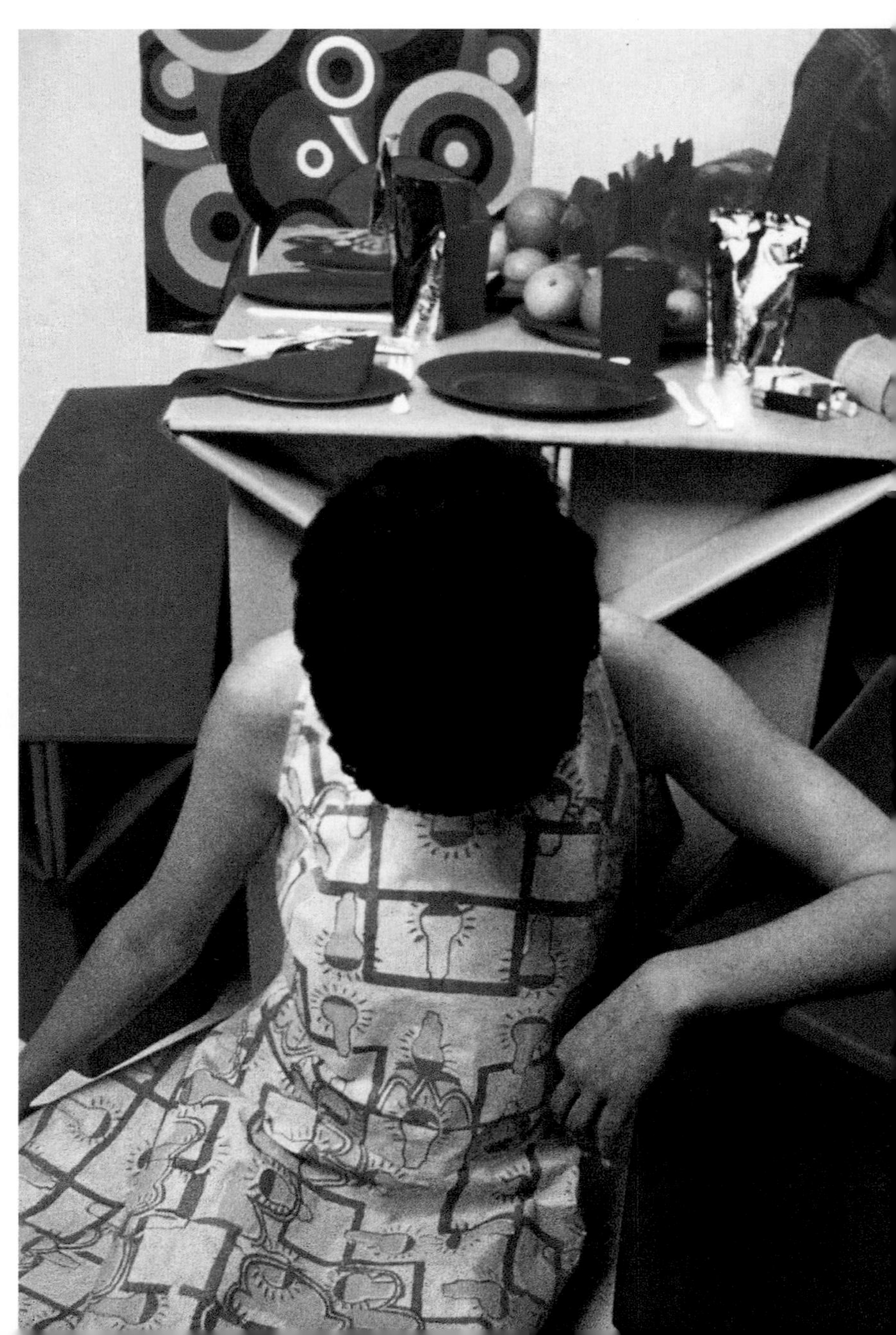

Above: Fashions from Mr Freedom, Tommy Roberts' boutique, found inspiration in the supermarket. OMO handbag, 1970. Right, top: Inflatable PVC furniture, the perfect solution for the pop environment. Right, bottom: Throwaway fashion in a disposable interior. Zandra Rhodes paper dress and paper shirt with paper plates and cardboard furniture.

wanted a bright inter-change of casual knits
found
in Tricel
Now have a whole mix and match wardrobe of striped and plain play-things in rich-looking high bulk Tricel.
It's all so easy now because Tricel knits always keep their shape, always look bright and young.
Left : Sleeveless sweater in nine stripe combinations style C231. All sweaters are about £4.9.6.
Bermuda shorts in six colours. Style C262. About 4½gns.
Above: Trousers, style C331X, in navy, yellow or four other colours. About 6gns. Teamed up navy sweater, style C233X, in four other colours.
Matching skirt, style C223, also available at about 3½gns.
Wide striped sweater, style C232X, four other colour schemes.
Susan Small IN TRICEL

Quant knits in Courtelle
Lee Target
27 super Courtelle patterns by Mary Quant
It's the rave look! The Quant look! High fashion in hand knitting. Mary Quant has designed a whole collection of knitting patterns for Courtelle. Because it knits slickly. Stays smart. Washes endlessly. Make this cosy sweater dress and socks in Lee Target Tarantelle. Pattern No. 6625 Get it in your knitting shop today!*
COURTELLE
A COURTAULDS FIBRE

FASHION IS EASY AS
A/B/C WITH Dacron*
Exotic culotte in DACRON polyester. Wash it often as you like. It dries in a trice. Style D969.
Combos of blue/mauve/lemon;
pink/lilac/cerise; lemon/green/pink.
Sizes 10-16–about 17 gns
by SUSAN SMALL
For nearest stockist write to:—
Susan Small Ltd, 76 Wells St., W.1.
DU PONT
*DACRON is Du Pont's registered trademark for its polyester fibre.
Du Pont makes fibres, not garments.
DU PONT CO (UK) LTD DU PONT HOUSE FETTER LANE LONDON EC4

This page: Fibre competition accelerated during the sixties as each of the main manufacturers attempted to couple their brand with the young and the fashionable. 'Tricel' (triacetate), made by British Celanese, competed with rivals Du Pont's Dacron (polyester) and Courtauld's Courtelle (acrylic).
Opposite: Synthetics made ideal 'fun furs' for sixties trendsetters. Fashion abandoned the traditional status symbols of the rich and flaunted fakeness.

Lock up
your
sons!

They don't stand a chance with 'Judy' almost-girls-about-town looking so terrific in their suits-and dresses-and trouser suits........
'Harv'—he's the one in the stocks—thinks Nicola's dresses are just great, (and really she thinks he's fab!)
Style No: 210007 machine-washable Judy dress in 'Crimplene' jersey. Navy/Pink, Pink/Navy, and Green/Navy. Sizes: 32"-38". Price about 4 gns.
Available at all leading shops and stores.
Judy
Judy
suits and dresses
for the very young; for the fives-to-teens and for the almost-girl-about-town.
IN
'Crimplene'
'TERYLENE'
By English Sewing Limited (Knitting Division) P.O. Box 5. Congleton, Cheshire. London Office 21 Cavendish Place, W.1. LANgham 2689

'TERYLENE': HIGH STYLE, SNUG FIT

Didn't want Janet to see me, sure to return ring if found out accidentally engaged to Diana night before. Bound to be scene. Thought cornered, as narrow alley, no way out. Stroke of luck, wearing suit in 'Terylene'. Wouldn't get torn, frayed, generally creased if squeezed self into narrow opening high up wall. So did so. Could sponge down after.

SUIT IN 55% 'TERYLENE'/45% WOOL, WORSTED.

'TERYLENE' IS THE REGISTERED TRADEMARK FOR THE POLYESTER FIBRE MADE BY IMPERIAL CHEMICAL INDUSTRIES LIMITED, LONDON

Above: Nylon bikinis in an unlikely setting. ICI published several magazines, including *Nylon Magazine* and *BNS Magazine* (British Nylon Spinners).
Above, left: Crimplene at its mid-sixties fashion peak.
Left: Sixties man in a sharp Terylene suit. ICI advertisement published in *Town* magazine.

The great new message was that art could be enjoyable, ironic and accessible, just as Marcel Duchamp had predicted: it could be anything, and so could fashion and design. Why not wear brash, shiny PVC, fake leather, 'fun' furs, glitzy Lurex, contoured Crimplene, heat-set acrylic slacks or sweaters made from plastic wool. Why not sit on expanded polyurethane foam made into 'fun furniture' or instantly furnish a room with transparent inflatable plastic and bright injection-moulded or cardboard tables and chairs. Pop culture bred a wave of new publications like the underground papers *It* and *Oz* and the more mainstream pop music magazines *New Musical Express* and *Melody Maker* which acted like manifestos for the new, young, music/art culture and where anything from a polyester 'Beatle' jacket, a nylon 'Beatle' wig or plastic elasticated 'Chelsea' boots could be bought by mail order. In the traditional press, the first newspaper colour supplements illustrated the work of new designers and fashion photographers and created a sense of excitement around the remodelled British culture.

Designers revelled in the ironic displacement of commercial art and in the playful Disney-fantasy world of man-made materials. It was a do-it-yourself wonderland of colour, texture and illusion. The designer Zandra Rhodes with her fellow art-school graduate Alex MacIntyre made their first home into a 'Pop environment, a perfect world of plastic, true to itself, honestly artificial'. Everything was covered in yards and yards of plastic: 'including the walls and even the television set. We had plastic grass carpets, collected plastic flowers and trees, used synthetic marble and Fablon tiles. We made the furniture by drawing shapes on the floor and building laminate-and-foam seating and tables on those spots. We built standard lamps from pillars of plastic and circled them with neon bulbs.'[22]

Up to this point, synthetic fibres were largely used in substitute roles – in the 1950s, nylon as an imitation of mink for coats or wool for carpets – but in the early 1960s their very 'fakeness' was celebrated as the young turned away from the wearing of unethical and expensive real furs and animal skins. The message was one of rejection – out would go the traditional symbols of luxury and in would come cheap plastic furs, dyed a vivid pink to emphasize their unnatural source. A whole new value system was being invented and the materials, forms and colours of elitist fashions

Zandra Rhodes wearing an Ossie Clark PVC suit and a James Wedge medallion helmet. Her companion is Alex MacIntyre.

were overthrown and replaced by innovative alternatives. London's young designers took up the new synthetic materials with unparalleled enthusiasm and with them revolutionized the silhouettes and structures of fashion making. Polyester double jersey was rediscovered as a youthful, relatively new design fabric; it could be moulded, shaped and used in ways that were free from convention. As has been noted in earlier chapters, social status is often reflected in consumer choice of fabrics. Naturals, like linen and silk, require expensive maintenance whereas the synthetics introduced a range of totally new clothing qualities which truly came into their own during the sixties pop era. Non-iron, permanently pleated, drip-dry, lightweight, easy-care, high strength and, above all, low cost fibres made synthetic fabrics the boon of high-street high fashion.

ICI was constantly experimenting with varieties of textured surface effects, with stretch, Bri-Nylon fabrics, leather-look simplex cloths, nylon chiffons and new foambacks for casual jackets. Its Textile Development Department commissioned numerous

Above: Professor Janey Ironside at the Royal College of Art with Lawrence Willcocks and his prizewinning menswear design, 1965. Opposite: Design in diagonally ribbed turquoise Crimplene by RCA student Birgitta Back for ICI, 1967.

designers to produce sample fabrics which were shown to its clients. It invested heavily in conspicuous links with the fashion industry and in particular tried to encourage tomorrow's leading designers to experiment with synthetic fabrics like Crimplene, Terylene, Bri-Nylon and PVC. To this end, ICI set up a series of projects within the fashion school at the Royal College of Art in the mid-sixties. In 1964, the men's outfitters, Hepworths, had celebrated its centenary by presenting the RCA with £20,000 to establish a menswear department in the school and ICI's competition to 'design leisure clothes for the future' gave these postgraduate students considerable scope for invention. Their designs ensured ICI's synthetics a young-fashion image – even the now legendary brand, Crimplene.

It is hard now to imagine the fashion euphoria that once surrounded Crimplene, the fibre success story of the sixties before it was killed off by over-production and over-marketing. Many people took it to be a completely new fabric but it was actually just another form of polyester developed by ICI in 1955. Its uniqueness lay in the fact that it had been heat 'crimped' – it was a polyester with a perm. Made into heavily textured fabrics, it was very difficult to cut and sew and was really only suitable for the shift dresses and simple lines that were, fortunately, fashionable at the time. Crimplene was an effective substitute for the expensive woollen jackets, dresses and coats more traditionally made from warp-knitted jersey. There is no better example of the commercial boom and bust profile of synthetic fibres than Crimplene. Once disillusionment has set in it is difficult to dispel a negative attitude to a fabric and within just a few years Crimplene's image became seriously tarnished and its magic waned. Killed off by over-exposure, it tumbled in status and became another synthetic laughingstock, forever haunted by the mass-market cliché of matronly boxy frocks worn with tight perms and accessorized with oversized plastic handbags. The rot was setting in for man-mades.

In the decade following the war, synthetic fabrics had been welcomed on the basis of their convenience but the 1960s was their first great fashion moment. Designers chose man-made materials not just because they were sponsored by fibre producers but because they best expressed the spirit of the age. The establishment equated mass-market materials with inferiority, poor quality, inauthenticity and above all bad taste, but the radical and forward-looking young designers who were the products of London's boom in art and design education took plastics and synthetics to their hearts. This was the first youth generation hat made clothes, music, graphics and furniture for themselves. 'Now nylon, Ban-lon, Terylene, Dacron, Crimplene, Corfam and a host of other names as non-natural as the materials they represent, by-products of unpronounceable chemical formulas, have ousted the "natural" fibres which until today composed the only coverings known to man', pronounced Madge Garland at the end of the decade.[23] Design was the magic wand that added value to plastics and synthetics and they became conspicuously fashionable on the streets of swinging London. The 'youthquake' was only possible because of plastics and synthetics and perhaps the last word should go to Janey Ironside whose students carried her philosophy into the world fashion scene: 'let us hope that the young everywhere will continue to shock and stimulate us, because that is the only way real fashion evolution will progress'.[24]

6

1970–80

The Disco Dacron Decade

'An American biochemical company claims to have isolated a substance that may be the key to human sexual attraction, a chemical that could induce females to fall for the most repulsive-looking men – even those wearing shell suits or flared trousers.'

Observer, 31 May 1992

John Travolta's portrayal of the disco prince of polyester in the ultimate seventies film *Saturday Night Fever* (1978) defines the moment when a drastic fashion collapse first hit synthetics. A disco dancing idol, clad in a white Dacron flared leisure suit and a black Qiana nylon shirt, his wardrobe spoke volumes about the showy but tacky clothes which finally led to the downfall of the fifties 'miracle fabric'. In just over twenty years, polyester, in all its brand guises, had plummeted socially and acquired the bad name that time has proved all too difficult to blot out. By the late seventies, polyester was stubbornly linked in the public psyche to pop-kitsch Abba performances or the Las Vegas bad taste of Elvis in a garish rhinestone jumpsuit. Worst of all, millions of shops were being flooded with rail upon rail of lurid rainbow-coloured shell suits – destined to become the sad epitaph of polyester and the most ridiculed of garments.
As recently as 1992, Roy Rivenburg reported that polyester still 'seems unable to shake its reputation as the clammy, sweaty and static-prone fabric of the disco era, and to counter such sentiment, manufacturers have tried everything from multimillion dollar PR blitzes to begging the federal government to change polyester's name – all to no avail'.[1]

Far Too Much of a Good Thing

The very proliferation of polyester was its downfall. Like nylon, it was everywhere and in everything; from underwear, to socks, frocks, shirts, carpets, curtains and bedsheets. In 1970, *Du Pont Magazine* was proud to publicize the new 'fake furs' then being produced by Borg and made from a knitted fabric that was originally developed for use in car-polishing pads and paint rollers. These deep-pile fabrics were 'generating genuine international high-fashion excitement' and

Opposite: John Travolta, the archetypal seventies disco idol in *Saturday Night Fever*, wearing a white polyester suit and Qiana nylon shirt.

Above: Glam Rock, glitter, platforms, shiny synthetics and the Swedish pop group, Abba – the essence of pop kitsch.
Below: Teen magazines radiate a taste for tacky fashions.

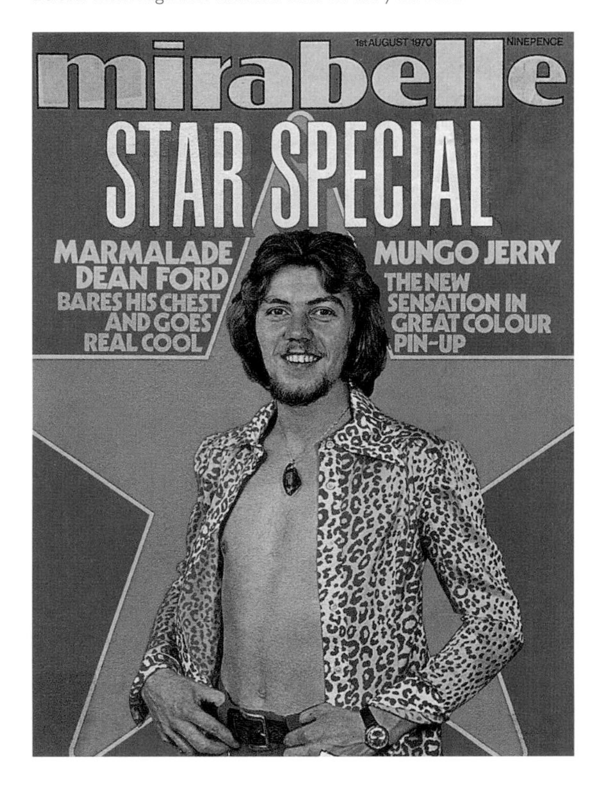

were earning 'glowing responses from sophisticated European high-fashion audiences' at a special preview in Munich. Especially for that occasion, John Boughton (a noted American couture furrier) had designed 20 women's and eight men's 'furs' made from paint-roller fabric and, in a nicely ironic twist, had trimmed them all with expensive animal furs. In 1968 Borg sold 14 million yards [13 million metres] of this fabric for a record profit of $38 million. 'The dramatic increases in these sales [were] explains in one word: fashion.'[2] This striking example shows the dangerous blind-spot that had developed in fibre-producers' eyes: quantity was everything, regardless of the quality or long-term merits of the fabric.

For years, convenience and the illusion of luxury had clouded the fashion consumer's viewpoint, obscuring the downside of synthetics – the dread combination of static, perspiration and sliminess next to the skin. Suddenly, it seemed, everyone was waking up to these physical discomforts and beginning to voice their criticisms of all the man-made fibres. 'Better dressing through chemistry' was proving to be a myth and disillusionment was taking hold as the advantages of 'wrinkle-free, dirt-repellent' clothes came to be far outweighed by the reality of the mass produced, cut-price fashions that synthetics had spawned. The 1970s is the decade that is most plundered when an outfit is needed to raise a laugh. It has become synonymous with countless fashion disasters: Orlon tank tops, polyester hot pants, nylon afro wigs, cow's-lick collars, sateen shirts and outsize Crimplene tunics. Out of the seventies came a new role for synthetics in social satire – as fibres of fun. '"The leisure suit was the height of polyester's success...but also the abyss", said Larry Hotz, a spokesman for Donna Karan. The idea of a comfortable wash-and-wear style suit was groundbreaking and lured many buyers, but it didn't take long for the jokes to start. "The fabric was unbearable," Hotz said. "It looked like it had a ton of starch in it and the colors (lurid lime greens, otherworldly oranges) were grotesque." Workmanship was another problem. Some manufacturers began cutting corners, flooding the market with polyesters that held up poorly and looked cheap.'[3]

The clothing manufacturing boom had created its own fashion Frankenstein and the glut of goods that were being produced and consumed was made from a seemingly limitless source of inexpensive synthetic textiles. Textile statistics show the leap in demand for textile fibres between 1950 and 1966, with by far the lion's share going to synthetics. 'The world production of all textile fibres increased from 9.4 million tons to 17.7 million tons, an increase of 4% per year compared to an annual rate of population growth of 1.7%. The synthetics' share of the market doubled. Silk had already lost ground to rayon and nylon. Cotton production increased, but its share of world fibres fell from 75% in 1940–41 to 49% in 1976–77. Similarly, wool, whose output remained stable in terms of quantity, saw its market share decline from 12.4% to 5.7% in the same period.... There was an eightfold rise in the production of man-mades between 1966 and 1976.'[4]

Above: A 'fake fur' coat made in 1970 from a Du Pont fabric designed for the car industry, lined with Qiana nylon, trimmed with real fur.
Below: The brief trend for hot pants was a favourite press topic.

stern magazin

Der erste Sexreport aus Afrika

Warum 10 Starfighter-Piloten ihren Dienst quittieren

Wehe den Geschiedenen

Wer keinen Beruf hat, sollte künftig nicht heiraten

Wohin rutscht der Rocksaum?

Wer die vorgeschriebene Rocklänge ernst nimmt, ist selber schuld

The Consumer Turn-off

More textile demand meant that more clothes were being manufactured and bought. Commenting on the immoderate and oddly dismal consumption of the seventies, Jeremy Seabrook wrote that 'the old coercive forces of poverty and want' had been replaced by the new tyranny of an 'insatiable and joyless consumerism'.[5] Synthetics had fostered the throwaway society and fed the new mass culture of materialism, but early signs that the bubble was soon to burst were already filtering into mainstream society by the late sixties. The 'shop your way to happiness' mood of the post-war years was already undermined by the seeping alternative values of hippy culture. Pop culture had shifted from gritty urban London to downbeat, free-spirited Southern California, where Timothy Leary's call to 'drop out, tune in and turn on' invited young people to turn their backs on society's standard values. Instead, they were to experiment with drugs, lifestyles, relationships and even clothes. This philosophy was positively un-American because one of its most important ideologies was the abandonment of the sacred ritual of endless consumption. During the Californian 'Summer of Love' of 1967, the media introduced 'hippy' culture to the rest of the world. There have always been drop-outs, simple-life movements and colonies of romantic socialists, but this was different; here was a cohesive anti-materialist movement on a near global scale that was driven by young adults. 'Naturalness' was all part of the hippy ethos and this applied to everything, from wholefoods to natural fibres, dyes and unstructured clothing. Once considered cranky, many of the attitudes rooted in this sixties counter-culture were the ones adopted by fledgling ecologists in the aftermath of the oil pollution and environmental scandals to come.

Already surfacing was the widespread public disenchantment with science and technology that found an early expression in *The Graduate*, a film made in 1968 starring Dustin Hoffman. It resonates with clashing values between affluent, complacent parents and their alienated children. Jeffrey Meikle described the powerful emblematic significance of the word 'plastic' in the film's opening sequence, when the young Dustin Hoffman is receiving advice on the secrets of life and worldly success from his father's best friend: '"I just want to say one word to you. Just one word: Plastics.... There's a great future in plastics." This odd pronouncement, isolated in the film's opening scene, convulsed audiences and became a line repeated into classicdom by a whole generation of kids. Most viewers would have had trouble explaining their laughter. Some perceived a comment on the banality of business, others an attack on comfortable middle-class materialism.... A few, catching an ominous note, entertained fleeting thoughts of science fiction nightmares, of technology run amok. And some merely relished the absurd elevation of the commonplace. Whatever the reasons, the scene hit a nerve and entered communal memory.'[6] Hollywood had given post-war Europe its first glimpses of the joys of consumer culture in the 1940s and 1950s. *The Graduate* shattered the seductive American Dream with its blatant contempt for conventional materialism and, in the process, made a joke of man-made materials in all their forms.

NUPRON.
MODULIZED™ RAYON
Charles Santore
Nupron Modulized rayon lives! It does the knit bit like it's never been done before. It breathes life into blends, textures and colors. It inspires freedom in styling and dressing. It gives comfort and carefreeness wherever it goes. In short, it promises you the world. So don't let it pass you by. C'mon out in Nupron Modulized rayon. Live like you've never lived before.

The Monsters of Science and Industry

It is the prevailing cultural attitudes to technology and nature that colour public acceptance of man-made materials and the positive image previously enjoyed by chemically created products was further blighted by the series of ecological disasters that occurred in the early 1970s. Massive oil spills in the oceans and nature-threatening chemical pollutants on land alienated consumers from the oil and chemical industries and stigmatized man-mades. Aldous Huxley's *Brave New World*[7], first published in 1932, found a new, young, popular audience at this time. His was the ultimate science-made distopia, a reminder to readers that every well-intentioned discovery in pure science is potentially subversive and threatening to humanity. Written at a time when Europe was undergoing massive industrial expansion and importing Henry Ford's dehumanized system of factory assembly lines, Huxley interpreted technology and mass production as the enemies of individualism and of humanity itself. In his projected future world, conformity and uniformity were the evil price of progress. Everything, from human beings to emotions and sensations, were synthetically simulated or chemically created. Families were extinct, test tube babies were manufactured in the Social Predestination Room and the drug Soma induced permanent passivity. Biologically engineered ageless human perfection and tedious uniformity were the predicted consequences of technological progress. Even in this universe of mass produced happiness, however, the social class system was still perpetuated and fixed and genetically standardized social roles were chemically blended – the Alphas and the Deltas, the leisured and the working classes.
In deference to the inventor of the twentieth century's brave new technologically made society, the single

surviving religion in the book, that of mass produced happiness, was Fordism.

In a world economy dominated by the ideology of ever increasing mass production, it was vital to maintain the momentum of consumption since the economy depended upon it, but Huxley had foreseen a time when humanity itself would become enslaved by its own addiction to industrial consumption, drained of its diversity and controlled by unseen economic forces. This was just the scenario that made Huxley's predictions seem so relevant to the anti-plastic movement of the 1970s. The growing psychological reaction to all things man-made was fuelled by the perception that technology had created too many standardized products for a mass society. Duplication and replication added to an underlying sense of social discomfort – the recurring vision of a chain-store polyester-clad nation living in prefabricated high-rise public housing. It was a nightmare of a caste-based uniformity, of Huxley's Deltas appropriately clothed in polyester shell suits in stark contrast to the Armani-dressed Alphas.

Over production without substantial investment in fibre development and improvement was building up to a classic boom and bust scenario. The market was being saturated with more and more synthetic fibres which had, by this time, begun to lose their novelty value. The initial euphoria that had greeted the wonder fabrics – acetates, acrylics, polyamides and polyesters – melted away as the uncomfortable nature of wash-and-wear became apparent. Bri-Nylon shirts lost their fashion cachet as they yellowed and sealed in perspiration, Orlon sweaters sparked and crackled with static electricity, and the credibility of synthetics went into reverse.

Crimplene, the great British polyester success story that had made a fortune for ICI in the sixties, crashed dramatically in the early seventies, largely because of unchecked over-production. It had become a high-street cliché, made into millions of rigid box-shaped dresses in tired shades of turquoise, pink or apricot. Not surprisingly, this led to the brand's total downfall. 'Crimplene became associated with this very limited range of fabrics which may have been fashionable – although nobody now wants to admit that they may have been eighteen and wearing a crimplene dress but somebody did, somewhere. The trouble was that it was coming out by the mile. Having got into chain stores in a big way (and,

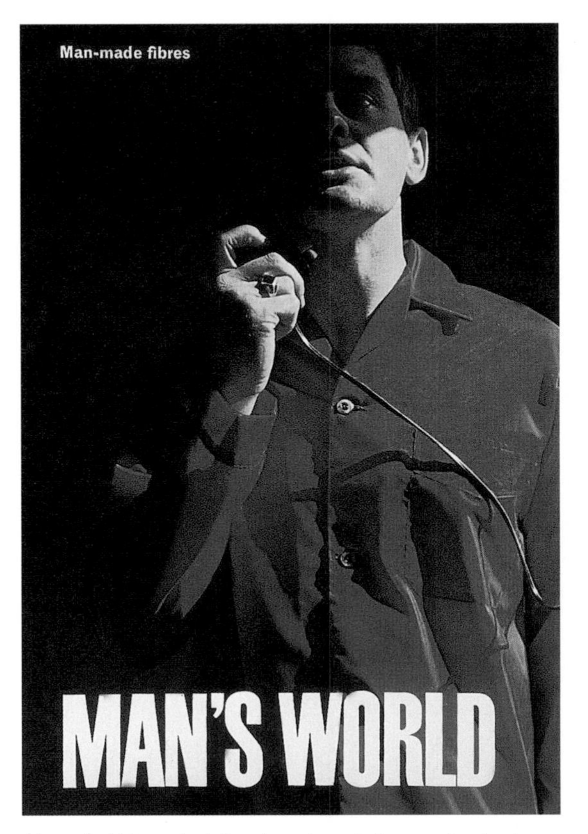

Above: A nightmare in static: nylon pyjamas between nylon sheets. Opposite: 'Back to Nature' in fibres, fabrics and lifestyles. 'Nupron' rayon, a semi-synthetic made from organic materials, identifies itself with the new 'natural' trend.

obviously, ICI wanted to go for volume) it suddenly became everything that your mum and your grand-ma were wearing and whatever designers were put behind it, it got a very middle-aged to elderly image and that killed it…its a great pity but it could never really recover – it was just marketed to death.'[8]

Courting the Home Dressmaker

It could well be argued that the synthetics industries had become dangerously complacent but, as the seventies progressed, a series of disasters began to pile up against them. In addition to a growing ecological consciousness among consumers, there was the global economic earthquake triggered in 1973 by the Yom Kippur War. Soaring oil prices saw a leap in the cost of fibre-producers' raw materials and, for their immediate and consumer customers alike, brought about a general recession. By 1977 desperate attempts were being made to shore up the fashion image of polyester by the two major fibre producers, Britain's ICI and America's Du Pont. 'High Fashion Means Big

Above: ICI enlists the aid of Mary Quant to rid Crimplene of its less than fashionable reputation. Quant designs were translated into Vogue and McCall's patterns for the home dressmaker in 1975.
Opposite: In 1977, McCall's patterns introduced the Infinite Dress which could be worn in 100 different ways. This one-piece, one-seam wardrobe 'should be made in Qiana nylon or Dacron polyester'.

Business', headlined an article in the *ICI Magazine* in 1976, describing their policy of encouraging top British designers to include chemical yarns in their ranges. In a bid to rescue Crimplene, ICI had, during the previous year, announced a collaboration with London's most famous designer, Mary Quant, 'the face that means fashion...with designs that have caught and expressed the tastes of today and tomorrow.'[9] ICI was looking to expand its market and what better way than to cut out all the middle-men and go straight to the home dressmaker. If she could be persuaded to choose synthetic fabrics then their future was more secure. Once again the familiar formula of linking fabrics with a well known designer was the first resort. 'Cut these out in Quant style!' shouted the magazine feature, which included the patterns and illustrated the 60 different polyester fabrics designed by Mary Quant especially for the home dressmaker. These double-jersey blends with Crimplene and spun Terylene were on sale only in Britain and appeared in more than 500 department stores and specialist fabric shops at £2 to £4 per yard (90 cm). Throughout the country special promotions were arranged at over 60 stores to introduce the new range. A follow-up Quant range was produced in the autumn of 1975, encouraging the home dressmaker to 'blossom out' by suggesting many layers of co-ordinating fabrics. 'A home dressmaker can thus produce a complete "layered" look: trousers, skirts, jackets, shirts, slipovers, evening dresses, overdresses, coats and capes' – all of which, of course, would greatly increase the volume of polyester fabric sales.

Endorsement from home-dressmaking pattern companies could make a very significant impact on fabric choices and company profits and specific synthetic textiles were recommended for each design. In 1976, the American McCall's Pattern Company entered into an agreement with Lydia of London, the designer of the Infinite Dress which was recommended to be made in Du Pont's Qiana nylon or Dacron polyester. Described as perhaps 'the most unusual and innovative garment pattern ever issued by McCall's', the 50 million 'home couturiers' in America were introduced to this 'wondrous one-piece, one-seam wardrobe in a single garment'. Enclosed with the pattern was a photographic sheet illustrating just 13 of the ways in which the dress could be worn (the designer was said to have discovered 100 ways).

Pursuing Control through Prediction

The desire to maximize profits demanded a careful balancing act from the fibre producers, between the mail order and chain store business where volume sales were made and the small firms who were closer to the cutting edge of fashion. Much of the trouble that befell polyester was caused by its success in the volume clothing market which labelled it as the fabric of cheap mass market garments. The one way out of this was to add value through design and major fibre producers became a formative influence in refining the profession of fashion forecasting. 'ICI Fibres and The Crystal Ball of Fashion', a report written for ICI in 1978, described the complex and difficult process of predicting the fashions of two years hence – and the importance to ICI Fibres of getting the predictions just right. 'In the exceptionally difficult environment in which the synthetic fibres manufacturers have been operating in the mid-1970s, getting the market right –

and helping downstream customers get it right, too – has become more important than ever. Mistakes mean clothes left unsold on retailers' racks: the blockage is felt all the way back up the pipeline as each connective part suffers a downturn in business.'[10] Proving the worth of prediction, the report expressed the view that 'John Travolta (in the films *Saturday Night Fever* and *Grease*) and the disco revolution will make their mark on fashion'.

Forecasting methods were becoming increasingly sophisticated and ICI, along with the world's other large fibre producers, were buying in the services of specialist design forecasters and fabric and colour consultants to help them identify and prepare for the major design shifts in textiles and fashion. The photographer Elsbeth Juda, who, with her husband Hans, had established and edited the elegant *Ambassador* magazine, was appointed by ICI to head a studio of young designers who would research and predict design trends and fabric developments. Based in London and working in conjunction with ICI's Textile Centre in Harrogate, this unit later became the highly respected Deryck Healey International (DHI) and concentrated on colour, fabric and garment-styling trends. Opinions were also gathered from French and Italian consultancies, from the latter primarily for knitwear in which Italy led the way.

ICI Fibres' Merchandising Department took great pains to analyze the fibres, colours and texture of proposed fabrics. It promoted products not only to the company's direct customers but also all the way down the line to the consumer. A regular bulletin booklet, *ICI Predicts*, was taken around to spinners, knitters and weavers, followed soon after by fabric samples and instructions on how to make them. These samples were rapidly produced at Deryck Healey's studio, 'where fabric from designs conceived only hours before are realised'.[11] Passing on authoritative and accurate trend information was in itself good public relations while seeing a design produced in a mass-market fabric was good business. Aimed at garment manufacturers and retailers, ICI also mounted Fabric Focus, an exhibition of hundreds of different commercial fabrics produced by their main customers in Britain and on the Continent together with a few sample garments illustrating how the fabrics could be used. In 1978, ICI was congratulating itself on the extraordinary growth rate of its own branded polyester: its in-house magazine reported that

In 1975, ICI carried out a male make-over experiment. Here, the smart new models stand behind cut-outs of their former scruffy selves.

Terylene was spanning the world.[12] In celebrating the twentieth anniversary of polyester fibre, the article proclaimed that ICI were producing 50 million pounds (22.5 million kg) of Terylene a year – 'a rate of growth unprecedented in the short history of man-made fibres'. ICI were determined that Terylene should retain its identity and trade name throughout the manufacturing process which meant that they took responsibility for 'seeing that not only the raw fibre but the finished products in the shops lived up to the claims made for "Terylene"'. The fibre manufacturer had, therefore, to be able to forecast general fashion trends up to 12 or 18 months in advance, create national publicity campaigns and inform the retailer about their plans at least six months in advance of the selling season. All this effort, ICI noted, was towards 'the ultimate aim of offering the public the unbeatable combination of "Terylene" plus fashion'.[13] Despite all these efforts, the fibre manufacturers still faced a fashion and economic reversal.

Irresistible Forces Condemn Unnaturals

Although world economic and political changes may have initiated many problems for the multinational chemical companies, they faced an even more destabilizing force: plastics, synthetics and all 'unnatural' materials were simply going out of fashion. Symptomatic of this decline was the press release issued by Du Pont in 1977 under the headline 'Polyester: The misunderstood Fiber'. It opened with a revealing statement addressed to editors: 'There appears to be a great deal of confusion today about what polyester is or isn't. Few of us can look into our clothes closets without finding it in some form or another. Yet it's been maligned, condemned, shunned and generally put down like no other fiber/fabric product. Maybe this is the right time to ferret out the facts from the fiction'.[14]

Dacron, Du Pont's miraculous polyester, had become a bad-taste cliché, epitomized by the double knit leisure suit in pastel shades. Du Pont's analysis of the crisis in polyester identifies the exact cultural climate that ostracized synthetic materials and shows that the company was acutely aware of the work that would have to be done on the fibre's damaged image before polyester could recover socially. In the quarter century since its invention, polyester had become 'the number one apparel fiber. Nothing compares to it. It's inexpensive, versatile and readily available. Almost everyone uses or wears something made with polyester.... [It] is a lot more than just "that stuff" that goes into the shiny, baby blue leisure suit your husband used to wear to everything. It can also look like a fine wool, camel's hair flannel or any number of natural-look fabric types that are so much the rage today....most of us would be in a fix if it weren't for this remarkable fiber product.'[15] Recalling the sensational launch of Dacron when audiences would thrill at the sight of polyester-suited businessmen jumping into swimming pools to demonstrate the quick-drying, wrinkle resistance of the wonder fibre, the press release went on to point out the many lifestyle changes the fabric had brought about. Iron and ironing board sales went into a decline while washing machine sales increased. Mothers had more leisure time, umbrellas were left at home, and clothes packing was no longer a problem. 'Life was glorious and easy-care when wrapped in a patina of polyester....Men's clothing, women's clothing, children's clothing, clothes for dressing up, clothes for dressing down, sports apparel, outerwear, innerwear, polyester was everywhere. It was the fiber of the '50s

In 1971, Emilio Pucci, the famous Italian couturier, created these print designs and a pants/dress uniform for air hostesses in Du Pont's Qiana nylon.

Living proof that the seventies was the decade that restraint forgot. Plaid jacket and 'Angels Flight disco pants' in 100 per cent Dacron polyester, 1978.

In 1971, Du Pont courted London's bespoke tailors by sponsoring a competition, 'The View From Savile Row'. There were 14 winning designs for a '1975 business suit'.

that kept going strong right on into the '60s. And no polyester fabric was more popular than the ubiquitous doubleknit that was in everybody's collection of leisure suits and pants suits.'[16]

Complacency and being out of step with the mood of the times were the factors that brought polyester into disrepute. Du Pont tried to counter this by promoting a new range of fabrics called 'Today's Dacron' which were said to combine the aesthetics of natural looking fabrics with easy-care and affordable prices, but not many believed it. Synthetics were sliding further and further out of the fashion sphere and Du Pont was falling back on stressing the practical advantages of synthetic fabrics – 'Wives Still Key Factor in What Husbands Wear' claimed one of its public relations statements in 1977.[17] Market Research had proved that despite the 'peacock revolution between 1970 and 1976, substantial increases were noted in the percentage share of female purchases of men's sport coats and suits'. In a survey it was found that 88 per cent of wives preferred to buy clothes as presents for their spouses. Females were said to have a 'built-in fashion antenna' plus a preference for practicality. However, in the same year Du Pont had to admit that 'the bloom was fading from the polyester doubleknit rose, especially in the eyes of a generation of flower children with their penchant for natural forms of everything. The environment was dirty: clean it up: politics were putrid: clean them up: food was crammed full of chemicals: clean them up: polyester was plastic and shiny: put down those doubleknits and put on the naturals.'[18]

At the beginning of the 1980s the 'back to nature' movement really kicked in, 'the granola-heads arrived and chemistry-set clothing fell further out of fashion. One by one, polyester manufacturers shut down plants, merged or dropped out of sight altogether. It was a blood bath.... The companies that survived had to shift focus. Eastman Kodak Co., creator of Kodel-brand fabric, turned to polyester-based plastic soda-pop bottles, cigarette filters and X-ray and camera film. Others used the substance to make artificial blood vessels, tire cords, pillows, sleeping bags and upholstery.'[19] All the press releases and promotional activities on behalf of synthetics in the late seventies had about as much effect as 'dripping water on a rock'.

Banished to the Blends

As the reputation of synthetics reached an unprecedented low, they were largely extinguished and banished to the anonymity of the blend. Increasingly, the sensitive word 'nylon' disappeared from garment labels and was replaced by its chemically correct alias 'polyamide'. Many consumers had no idea that nylon and polyamide were one and the same fibre and, although nylon had tumbled in status, it at least had a fabric identity; polyamide, on the other hand, was a term used in the laboratory that meant nothing to the consumer, so this name change did nothing to restore nylon's credibility. In fashion terms, chemical fibres were forced underground, into cotton and other blends, children's clothing and underwear. If synthetics existed at all in the middle or high market they were obscured as a small percentage in blends, intended to add background strength to the natural fibres. Comfort and luxury have always been synonymous with natural fibres and the manufacturers of the 'noble naturals' – silk, wool, cotton and linen – seized their moment to run vigorous promotional campaigns with evocative slogans like 'Wool Runs on Grass'. During the seventies and early eighties, fashion turned away from the shameless plasticity of the mid sixties and a 'back to nature' romanticism held sway. Mainstream fashion increasingly took its influences from the off-beat wear of hippy culture. Hippies had introduced ethnic clothing to sophisticated urban capitals. Widespread travel to India, Tibet, Turkey and Afghanistan flooded Western fashion with ethnic influences as well as with authentic costumes. The emphasis was on natural fabrics (and plenty of them) and on fashion which imitated the styles of ancient cultures, made using traditional materials and methods: cottons, wools, silks and linen, hand-spun and hand-woven, coloured with vegetable dyes, printed with woodblocks, hand-made and decorated with hand embroidery. Dress shapes were simplified and drawn from the traditions of joining together squares and oblongs of fabric, as though they had just been taken off a handloom. All this may have had little or nothing to do with authenticity, with the hard living of the peoples of less developed countries, but its homespun nature was all part of its attraction as fashion turned green. By the mid-seventies, the only form of synthetic that was overtly 'fashionable' was the black plastic bin liner appropriated by the punks – the self-confessed 'blank generation' who had chosen the one tacky material they were sure would incite the most public offence. PVC and rubber, the fabrics of fetish clothing, were also resurrected in this negative fashion context.

For retailers such as Laura Ashley, a romantic nostalgia for the English past laid the emphasis on traditional fabrics like printed voiles and classic weaves. Gathered skirts, smocking, frills, hand-block prints, layers, trimmings – the end of the man-made aesthetic was reflected in the wholesome excess of Laura Ashley's recreation of the simple country life. The 1980s saw the 'gentrification' of natural fibres and re-established the class barriers between fabrics. Ralph Lauren and Calvin Klein sold a massive amount of homespun, high-priced 'naturals' and Armani's fine wools and linens became the aspirational fabrics of the 'designer decade'. Naturals were the ultimate status fabrics as they required expensive maintenance while hinting at a long heritage of aristocratic tradition and quality. In direct contrast to the synthetics of the fifties and sixties which were equalizing fabrics, intended, initially at least, to provide good quality materials for everyone, natural fibres began to command exclusive, high prices. *American Fibres and Fabrics* were disturbed to record in 1985 a kind of 'class warfare', 'fuelled by the elitist and pretentious advertising of designers such as Ralph Lauren and Calvin Klein, hearkening back to a natural way of life that never existed. I am sure that most customers who followed Mr. Klein's view of "Back to Nature" would be surprised that the tiny Mexican village serving as the backdrop for his recent ad campaign is today still mired in abject poverty, with hardly any indoor plumbing.... Additionally, we were dismayed by so-called consumer research which led Cotton Incorporated to tell this publisher that wearing polyester may be "socially risky". From every point of view such attitudes are distressing.'[20]

Fashion is fabricated in cloth but shaped by opinions and more and more designers viewed synthetics with growing distaste; socially déclassé fabrics, they represented garments worn from necessity rather than choice. In 1984, *American Fibres and Fabrics* reflected upon the reasons for the fibre 'counter-revolution' of the previous decade and on the lingering loss of synthetic status. In an entire issue dedicated to polyester, it noted: 'When selecting the contents for this issue, we were surprised by the inflexible attitudes held by so many we talked to who should know better.' When, around 1970, an oversaturated market led producers to compete on price alone, there was 'fashion indigestion' in which stores were flooded with similar colours and textures and a massive homogenization of styling. This was when doubleknit became a by-word for bad taste, tarnishing polyester's name. But to dwell on this moment, the magazine argued, was unfair. In the interim, fibre engineers had been hard at work 'naturalizing' the synthetics to meet objections to their 'hand' or feel and creating new variations, demonstrating that solutions were in sight for even the breathability problem. The special issue included a section called 'Touch and Tell' with 15 fabric swatches and an answer key that allowed readers to test their ability to discriminate between natural, synthetic and blended cloth.[21]

A study published in 1994 reveals the continuing taboo against synthetic fabrics in the United States. 'Scores of college educated American women and men [are convinced that] polyester – the most versatile and emblematic of the synthetics – . . . does not 'breathe'; that it 'feels' inferior; that it comes in garish or less than subtle colours. [It] feels like Saran Wrap [clingfilm] on a hot day; provokes uncontrollable itching and sweating; is a 'yucky' plastic. . . . casual probing has also elicited numerous references to class stigma: the word polyester conjures up the image of a lower middle class tour group filing off a bus at Disneyland in pastel leisure suits.'[22]

The post-war synthetics industry collapsed under the weight of its own success. A combination of low cost, durability and design versatility made mass fashion available which ultimately resulted in the loss of synthetics' credibility and status. Made into

Above: A Bri-Nylon bikini, just the outfit for scaling scaffolding. Fibre advertisements had to be eye-catching.
Opposite: 'There is no future in England's dream/We are the flowers in the dustbin', from *God Save the Queen* by the Sex Pistols, 1977. The 'blank generation' turned the plastic binliner into a symbolic uniform.

millions of factory-produced identical multiples, synthetic clothes created a 'sameness' which was their downfall. Polyester, the emperor of all synthetic fibres, was overthrown in the 1970s and the peasant class of cottons and wools triumphed in the first ideological war in textile history. Silk was restored to its couture throne and the difficult-to-care-for linen became the emblem of the designer aristocrats of the next decade. The early eighties brought the 'High Noon' spectacle of Versace and Armani – a designer-label fightout in luxury fibres. But coming, it seemed, from nowhere, were the newest outlaws of fashion – Yohji Yamamoto, Rei Kawakubo and Issey Miyake. Rulebreakers from Japan, they invaded the West, armed with tradition, expressed through technology and laden with explosive intellectual concepts. They embraced synthetics and elevated them to an entirely new aesthetic plane.

7

'We must learn the essence of textiles and garments; that which makes them more than mere cloth and clothes and elevates them to the level of a cultural expression'

Junichi Arai

1980–

Japanese Design and the Fine Art of Technology

During the 1980s, Japan introduced Europe to the creations of fibre engineer-philosopher-designers which marked the beginning of a sophisticated and alternative design aesthetic for synthetics.
This paradoxical cocktail drew on Japan's ancient traditions of hand weaving and dyeing and combined these with futuristic fibre- and fabric-making technologies. Japan had engineered and designed itself out of a disastrous defeat in the war and, almost unnoticed, had eclipsed the rest of the world with its technological inventiveness. It is in Japan that 'fabric weaving' and 'engineering' are spoken of in the same breath and it was Japanese fashion designers who introduced the West to a completely new philosophy of clothing, in all its respects, from textile textures to vivid synthetic colours and bizarre silhouettes.

The Japanese are the masters of the marriage between twenty-first century technology and the ancient craft of textile making and their greatest fashion designers, such as Issey Miyake, Yohji Yamamoto and Rei Kawakubo, have had an unparalleled and direct involvement with fabric; with fibres, dyeing, texturing, finishing and pleating. Textile technologists and designers like Junichi Arai, Riko Sudo, Makiko Minagawa and Hiroshi Matsushita have helped make Japan a major influence on high fashion and their expert manipulation of the inherent qualities of synthetic fibres was a key factor in extricating man-made fabrics from their association with the lower end of mass-market clothing. It was Japan's remarkable conceptualist designers who reintroduced synthetics to the West in their new guise of high-status artistic fabrics and they did it by being actively involved in fibre and chemical technologies and by working in close collaboration with textile inventors and creators.

Opposite: Yohji Yamamoto, Lycra and net minidress and Day-Glo pink tank top and pants. Nadja Auermann for British *Vogue*, 1993. Photograph: Nick Knight.

Yohji Yamamoto advertising campaign, 1986. Photograph: Nick Knight, Art Director: Marc Ascoli.

The Conquest of High Fashion

Synthetic fibres suffered a serious image decline in the wake of the economic and ecological disasters of the 1970s. Disgraced, downgraded and stigmatized, by the early 1980s it was hard to imagine that nylon or polyester would ever again scale fashion's peaks and it seemed that synthetics had found their natural social level, down at the lower end of every high street. Just then, however, fashion had one of its periodic cultural shocks when Japanese designers invaded the Paris catwalks. In 1981, Rei Kawakubo and Yohji Yamamoto made fashion history by becoming the first Japanese designers to be invited to show at the prestigious Paris Collections. Before these sensational shows, one of the very few fashion certainties was that design was led by the West, usually by Paris, sometimes London, and occasionally by Milan or New York. The East had been a fashion influence but never a leader – not until Tokyo's philosophical and enigmatic designers overturned fundamental Western ideas about what fashion could be. By the mid-eighties, the asymmetrical cuts of Japan had swept through the fashion world like a sect, establishing a new holy order of geometric minimalism. Fashion design in the West is based on shaping, by cutting into cloth, while the Japanese aesthetic is essentially unstructured, respecting the integrity of the woven fabric, with design solutions owing much to the wrapping and tying techniques of Japan's national dress, the kimono. For the first time in the world of high fashion, the customer was actively engaged in the shaping of their own outfit. Gone were the armholes, darts, padding, bones, seams, buttons and zips – the 'this is the way to wear me' indicators – instead, an oddly shaped cloth had to be layered and wound around the body before the asymmetrical garment emerged. The clothing forms created by Japan's designers were a revelation in the West and many of these shapes were possible purely because their fabrics were so technologically advanced. While the best the West could do to revolutionize the eighties silhouette was to exaggerate the shoulder line with padding, Japan was parading strange, sculptural constructions before the world's astonished fashion press. There was no point of reference, it was neither 'power-dressing' nor 'street style', it was something else, a combination of art and science, far removed from couture and social status. Japan was a technological hothouse, both in its miniaturized electronics and in its microfibre technologies. It presented a jaded fashion world with something new and outside the European experience – clothes that were more connected to technology and artistic ideas than to couture and social status. A totally different clothing concept and textile aesthetic washed over the West, and, in Rei Kawakubo's early collections, fashion editors found their most lasting and prized dress code: a legacy of secure and eternal shades of black. Yamamoto, Miyake and Kawakubo are among the string of Japanese designers who are now fixed in the fashion vocabulary but a generation ago it would have been difficult to name even one.

Designing the Sense of the Times

Rei Kawakubo is one of the century's most innovative and influential fashion designers and is credited with exploding all the clichés and traditions of Western fashion. Born in Tokyo in 1942, she graduated from Keio University in 1964 with a degree in Aesthetics. Her transformation into a world famous designer started almost by chance when she joined the advertising department of a manufacturer of acrylic fibres. This brought her into contact with many fashion professionals and she soon became a stylist. Frustrated by the fact that she could not find the clothes she wanted to work with, she decided to design her own.

Since 1969, when she first introduced the Comme des Garçons label, Rei Kawakubo has held a unique position within the fashion industry as the torchbearer of the avant-garde. She has consistently subverted conventional ideas of how women and men should look, creating what are initially thought to be weird garments and bizarre silhouettes that, within a few years, have been absorbed in a diluted form into the fashion current. It is one of the great lessons of modern culture that truly lasting and original works are often alarming at first and even the most hardened of journalists and buyers are said to have been 'frightened' by some Comme collections. Kawakubo defines her career as 'designing the sense of the times' and her reputation for getting it just right is one that has been proved over the years.

Although she still sells most of her clothes in the East, it is in the West that Rei Kawakubo has made her reputation as one of the world's most creative designers.[1] Her designs are considered to be impenetrable by many but she is much admired as a designer's designer, the intellectual face of fashion. Her collections are uncompromising and complex technical distillations, too strong an elixir for the faint hearted but enthusiastically swallowed by the hardcore fashion buffs. Straddling the worlds of business, culture and fashion, Kawakubo's global enterprise has been described as 'a total work or art', encompassing fashion shows, famous photographic catalogues and shop installations, often created in collaboration with artists.

Rei Kawakubo challenges accepted perceptions of the beautiful and her early 'deconstructed' clothes have set up an entirely alternative aesthetic, one which finds elegance in the different – even the shockingly so. In the late seventies and early eighties when her torn, crumpled and, to Western eyes, shapeless 'ripped' garments were shown in Paris, she was criticized by the media and ridiculed by satirists. When British *Vogue* featured her controversial 1983 collection, an outraged reader demanded to know why anyone would want to pay $230 (£150) for 'a torn shroud'. Only Rei Kawakubo could have put three sleeves or two neck openings into a garment or

Rei Kawakubo, Comme des Garçons, 1986. Black polyester georgette with bonded polyurethane.

would have designed a sweater that looked as if the moths had practically devoured it. Clothes were specially woven with tears in the fabric but these were not holes but 'lace' and the seemingly random rips were carefully designed and technologically created to illustrate the beauty of imperfection.

'I start from zero', Kawakubo has said. Beginning with the thread itself and working outside the Western fashion tradition, she has rethought the entire construction of clothes and her close involvement with textile manufacturing processes has resulted in many new fabrics, some of which have looked as if they were bubbling and boiling or even melting over the body. Kawakubo has often used chemically abused cloth, sometimes with dip-dyed or bleached out sections; abstract shapes have been created by bonding cotton, rayon and polyurethane, and all manner of chemical techniques have been used to create the right effect – dévoré, flocking, heating, melting and texturing of polyester, and boiled, frayed wool that became such a mainstream fashion trend following one of her collections in the early 1990s.

The Magic in the Fabric

Japanese designers have shifted the whole emphasis of nineties fashion away from the cut and onto the cloth and behind all the famous fashion names are an equally talented list of cloth makers who have amazed the West with their ingenuity. Rei Kawakubo works in close collaboration with the Japanese textile company run by Hiroshi Matsushita and, describing their method of designing, he has said, 'My company does not come up with new ideas for textiles and simply present them to Rei Kawakubo for her to use in her designs. Instead, it begins as a much more vague process whereby Kawakubo reflects upon the last collection and then starts thinking of what she is looking for in the next collection. It may be a simple word like "light" that she uses to describe the fabrics she is interested in and that is where we begin. . . . both man-made and natural textiles are drawn upon in order to produce an effect, feeling or image. . . . We rarely repeat a fabric and each collection is composed of original fabrics made exclusively for that season.'[2]

Issey Miyake has worked in a creative collaboration with his textile director Makiko Minagawa for more than twenty years and between them they have produced countless pioneering fabrics and reinvented the relationship between the body and clothes. 'Flying saucer dresses' were made from twisted ropes of cloth and Miyake's legendary pleated fabrics remain both luminous in colour and permanently shaped thanks to polyester. In the early 1980s, Miyake designed the famous 'second skin' silicone bustiers that were shown in his 'Bodyworks' touring exhibition in Tokyo, Los Angeles, San Francisco and at London's Victoria and Albert Museum.

Experimentation is expensive and a Miyake garment often carries a four-figure price. At the end of the 1980s (just in time for the recession), Miyake started using an inexpensive, extra-lightweight permanently-pleated polyester. Anti-static, perspiration-absorbent and fast-drying, it could be hand or machine washed or dry cleaned and, above all, needed no ironing and folded into a tiny flat square – the ideal travelling companion. Miyake's famous 'Pleats Please' collections have since become his

Top: Rei Kawakubo, Comme des Garçons, 1995.
Centre: Rei Kawakubo, Comme des Garçons, 1997, poyester and paper.
Bottom: Junya Watanabe, 1996.
Opposite: Rei Kawakubo, Comme des Garçons, 1997.

Above: Issey Miyake, Tatoo Body, printed polyester tricot, 1989.
Opposite: Issey Miyake, 'Pleats', 1989.

design signature and are considered to be 'classic' garments. This was a genuinely modern clothing concept where the fabric could be contained in the most basic, two-dimensional shapes but could expand around even the largest three dimensional body. He established a range of nine colours that were available year-round and then kept adding pieces, all priced at more affordable levels, £100–300 ($150–500).

Metal, plastic, laminated paper, perspex, silicone, bamboo, rubber – to Miyake, anything can be turned into clothing. The relationship between the body and clothes is, it is clear, differently understood in Japanese culture. A kimono may be 'a shape frozen in time' but it has taught many designers the principle of using the space between the body and the cloth. The thermoplastic nature of synthetics makes it possible to preserve any shape, crinkle, pleat or crease in fabrics simply by heat-setting, moulding or bonding the fibres. *Time* magazine described Miyake's clothes in 1986, at the height of Japan's influence on fashion, as 'ancestral and futuristic all at once'; they are, it stated, 'declarations of independence for the body....like so many pieces of whole cloth finding fresh form in the controlled accident of the fall...making the body under them feel as loose and free as the fabric....His designs challenge so many traditional expectations and break so many rules that they need different sets of standards to be understood or even worn. "I know many people resist or reject my clothing because it's not a package that's already formed, like European clothing. Without the wearer's ingenuity, my clothing isn't clothing."'[3]

Modern synthetics have been given striking new powers of simulation in Japan. 'Make me a fabric that looks like poison', Miyake once asked Makiko Minagawa, and in the last decade we have seen fabrics that recreate shimmering liquids, melting metal or iridescent oil – cloth can now be technologically manipulated to visually replicate virtually any surface. Sheen and shimmer are built into Miyake's fibres and coats are cut using high electric charges which burn and bond the seams together. Polyester, shrunken nylon, microfibre taffetas and monofilament gauzes undergo unspecified chemical and heat treatments to emerge as textured cocoons for the body. Waves or wrinkles are technologically fixed into the fabrics or, conversely, creaseless, glassy acetates produce wrinkle-free clothing.

Proof that Japan's unique contribution to modern fashion and textile design will continue lies in the emergence of a new generation of designers who also specialize in fabric-based design. Two of the most notable and interesting in terms of synthetic fashion are Junya Watanabe and Yoshiki Hishinuma. Watanabe began working with Rei Kawakubo in 1984 and in 1992 launched his own line under the Comme des Garçons label and is now considered to be one of Japan's most influential younger designers. Like Watanabe, Yoshiki Hishinuma learnt from the best when he was an assistant at the Miyake Design Studio. In 1984 he started his own label in Japan and has shown a collection annually in Paris. His career embraces product design (including garments for Alessi), theatre costumes and exhibitions and he designs both textiles and clothes. However technical and sophisticated the processes, there is something of the hands-on, artisan feel about the Japanese approach to textile and fashion design and it is hardly surprising that the finished articles seem equally if not more at home in an art gallery as in a shop.

Flawless Imperfection

'What I want to make is not "bread" [profit] but "roses" [contentment]', wrote Junichi Arai in 1989 and, more than any other Japanese textile designer, he epitomizes the philosophy of blending the new with the old and the techniques of the past with the technologies of the future. He has created many textiles for fashion designers such as Issey Miyake, Yohji Yamamoto, Kansai Yamamoto, Hanae Mori and Rei Kawakubo and, more recently, has worked with Yoshiki Hishinuma, a designer he holds in high esteem because he 'fights with cloth with strong character, tames it and creates clothing'.[4] The opposite poles of low and high technology are drawn together in Arai's work, refuting the conventional idea that hand-crafted fabrics are distinctively individual and machine-made are impersonal. He has often said that 'the human hand and technology must not be separated', and his own work 'challenges the prejudice that machine-made products are by definition more lifeless than hand made products. He has proved that symbiosis between the old and the new, simplicity and complexity, the natural and man-made, handicraft and high-tech industry is possible.'[5] The secret is to produce a computer-programmed high-tech cloth that looks hand made.

The grandson of a yarnmaker and son of a kimono clothweaver, Arai revived many ancient weaving techniques in his development and pursuit of textile design and united them with new materials and high-tech computerized looms. His experiments have crossed the full range of textile technologies a nd he is respected worldwide. 'His boundless fantasy and unconventional approach led to unbelievable experiments in the fields of yarn making, weaving, knitting, yarn and cloth dyeing, finishing and pattern design. In doing so he makes use of all sorts of equipment: current, outmoded and the latest. There is hardly a material he hasn't used or treated. Not only the natural fibres, such as silk, wool, cotton, linen, paper, but also a vast array of the man-made fibres: rayon and cupro, acetate and tri-acetate, polyamides, acrylics, polyester, polypropylene, elastomerics, glass and metal fibres (including aluminium, titanium, copper, stainless steel).'[6]

By the late 1970s, when fabric development and fashion in the West had reached an impoverished low, Junichi Arai was already experimenting with computers and was soon producing textiles for Issey Miyake and Rei Kawakubo using computer-aided looms. Arai appreciates the most negative aspects of technological production – the blight of standardization and the bane of simulation. 'It's hard to find a genuine thing in life,' he has said, 'everything that we find in arts and crafts today is a perversion from the real thing to production motivated only by money...life today is rich and easy because of modern industry and technology, but individuality and uniqueness are sacrificed. Everything is standardized and has lost personality. This is the necessary evil of the modern world...our greatest goal is to balance technology and the human spirit. In order to clothe the modern man who dreams of a spiritually harmonized future, textiles, the basics of everyday life, must exclude the harshness of modern industry and be woven with the philosophy of tomorrow... Textiles worn so close to the skin and heart of man should be treasured sources of comfort...The future of textiles is in texture, and texture is infinite, it is what your fingertips feel and tell your heart.'[7]

Japanese design sees no incongruity in the fusion of science and romanticism, and as one tangible expression of this, Arai is working with technology to overcome its own in-built blight of uniformity. He believes that, as an ideal, fabrics should be custom-made for specific clothes and his dream is to realize 'the personal textile'. Using a computerized system that adjusts a design to make a unique jacquard patterned fabric in conjunction with a video record of the client's figure, Arai aims to transmit information to a cutting machine 'which cuts a selected dress style, which will be sewn for the ultimate in personal attire'.[8] A truly individual garment achieved through high technology.

Junichi Arai introduced new technologies to the old mills at Kiryu in north-central Japan and from there he produced fabrics to order for the major Japanese fashion houses. In 1984 a collaboration between Arai, Reiko Sudo and Michiko Tomidokoro was launched based on Arai's principle of 'direct from maker to user'. Known as the Nuno Corporation, they opened a shop in Tokyo's AXIS Design Centre. Arai stepped down as company president in 1987 and, since then, has concentrated on pursuing his own creative activities in art textiles and on providing technical and artistic guidance to a new generation of textile makers. His licensed product line is called 'Poetry in Fabrics'.

Junichi Arai, 1997.

Junichi Arai, 1997, 100 per cent polyester.

Junichi Arai, 1997, 100 per cent polyester.

Reiko Sudo, 'Stainless Series' for Nuno, 1990, 100 per cent polyester treated with powdered steel using the car industry's spatter-plating techniques.

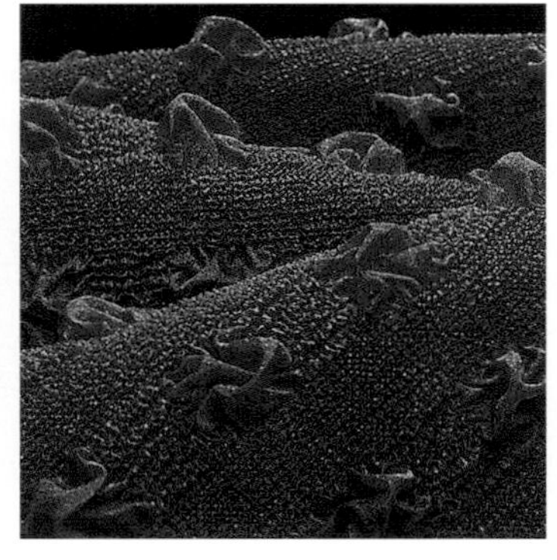

Reiko Sudo, 'Jellyfish' for Nuno, 1993, in 100 per cent polyester organdy, heat-set into crinkles using industrial vinyl polychloride fabric, developed for car-seat covers.

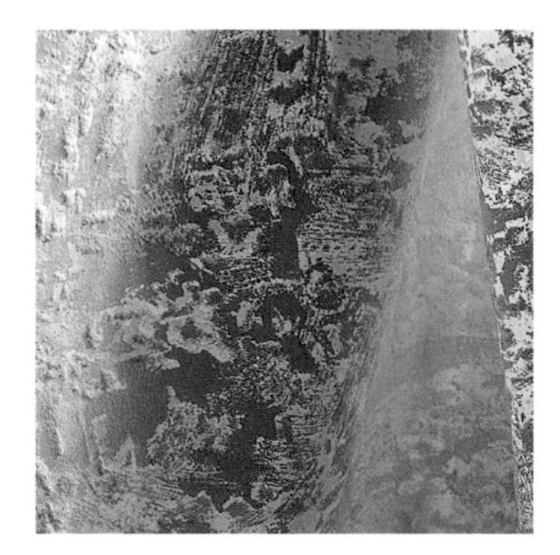

Reiko Sudo, 'Turkish Wall', manufactured by the Nuno Corporation in 1995 in 100 per cent polyester, using pigment-printing and flocking processes.

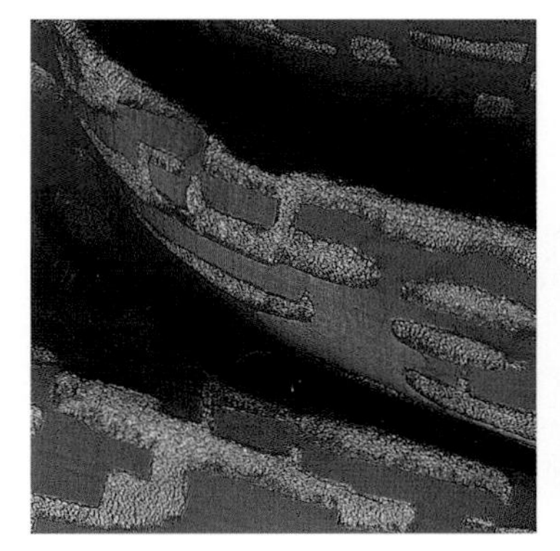

Reiko Sudo, 'Graphite' for Nuno, 1997, polyester and uncut rayon velvet piling.

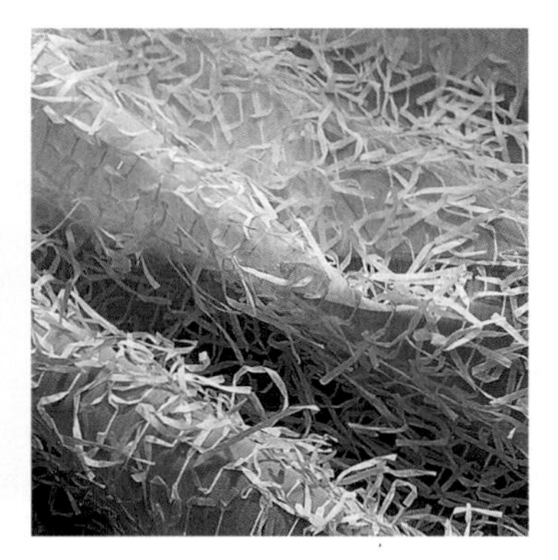

Reiko Sudo, 'Frazzled Paper' for Nuno, 1997, handmade paper and polyester.

Reiko Sudo, 'Combed Paper' for Nuno, 1997, 43 per cent polyester, 57 per cent paper. Strips of handmade paper are woven into a sheer polyester organdy base.

The Intelligent Hand Meets the Sensitive Machine

The Nuno Corporation continues to weave together art and science and its speciality is in developing new substances, processes and treatments for fabric making, such as 'crumbling', 'hand-ripped' and 'cracked' cloths and a fabric incorporating hand-inserted feathers. None of these techniques would be thought to be viable in conventional European textile-manufacturing industries yet Nuno textiles are widely produced in Japan. Continuous experimentation is the ethos of the Nuno company and, to quote Reiko Sudo, progress begins where 'the intelligent hand meets the sensitive machine'. 'In Japan, as elsewhere,' she writes, 'the advent of mass-produced, standardized fabrics made hand-weavers a luxury affordable to only a very few. And yet industry and handicraft did not have to be opposites; between factory and cottage existed a middle ground of untapped possibilities.' Nuno's award-winning textile designers have produced all manner of experimental fabrics: 'Threads with "incompatible" shrink ratios are intertwined and tossed into a hot dryer to yield unprecedented sculptural textures. Metallic films are bonded to conventional threads then melted to create transparent filigrees. Computer-programmed looms dance to digitally-enhanced African tribal patterns. Chemically-reprocessed Okinawan banana fibre-coated cottons complement laboriously hand-finished synthetics.'[9] But Nuno emphasize that their works are meant for 'mass production and normal use', not as exclusive 'craft' materials.[10]

Reiko Sudo adapts and converts technology and materials from unlikely industrial sources such as the electronics and automobile industries. 'Copper Cloth', a taffeta-type fabric which is malleable and holds any shape it is bent into, was introduced in 1993. The copper wefts of this fabric are the same material as is used in telephone lines; the polyurethane coating that guards against electric shocks and signal noises also prevents 'greening' and brittleness, making the material an ideal weaving thread. The warps are of 'Promix', a Japanese fibre created from casein (a milk protein) and acrylonitrile. The construction of the cloth is on a base of 100 per cent wool with an overlay of 100 per cent rayon. The combination of copper and white fibre produces an iridescent pink textile. Such imaginative creations and combinations are what make Japanese fabrics uniquely innovative. A 1993 *New York Times* article on the 'New Era in Design Materials' referred to another of Sudo's fabrics which takes its effects from car manufacturing: 'In the Japanese automobile industry, three powdered metals – chromium, nickel and iron, the components of stainless steel – are blended into a liquid and splatter-plated onto the window frames or underside of a car to give it a stainless steel finish. Ms Sudo said she applied the same technique to cloth for a fabric that reflects light.'[11] Manufactured by the Kanebo Corporation, 'Stainless Steel Emboss', 'Stainless Steel Pleats' and 'Stainless Steel Gloss' were difficult fabrics to realize as early attempts to evenly cover a large, flat, flexible surface such as a polyester fabric using a process designed for doorknobs and other small, solid items presented many technical problems; once these were overcome, a viable cloth could be made. Other fabrics by Reiko Sudo have been made from recycled PET bottles coated with polyurethane resin.

Attitudes to age and time in Japan are quite different from those in the West and many of Japan's designers appreciate and respect the aging of cloth as much as that of the person. Yohji Yamamoto believes that 'cotton is alive – it needs time, if you want the real touch of it you have to wait for 10 years'. This philosophy was made literal in Nuno's chemically-created 'Ragged Series' which was inspired by patches, repairs and worn areas in old kimonos and mattress covers and which resulted in fabrics that were collages of different materials of different ages. These aged materials tell us, in Reiko Sudo's words, 'the history of a person's life and the colours of the periods they lived through. That is beautiful – real Japanese culture.'[12]

By showing the West an alternative way of dressing which depends on the inherent qualities of synthetic fabrics, Japan has reversed the status decline of man-made materials. A complete economic inversion has taken place in the fashion value of textiles. A Japanese polyester or rayon may now be priced at £300 ($500) per metre, whereas silk is all too often a low cost, third-world fabric. This fact clearly shows that value is not inherent in the fibres themselves but is created for fabrics by association with quality design, technological innovation and consumer opinions, as well as successful promotional campaigns.

Opposite, top and centre: Yoshiki Hishinuma, 1995.
Opposite, bottom: Yoshiki Hishinuma, 1997.

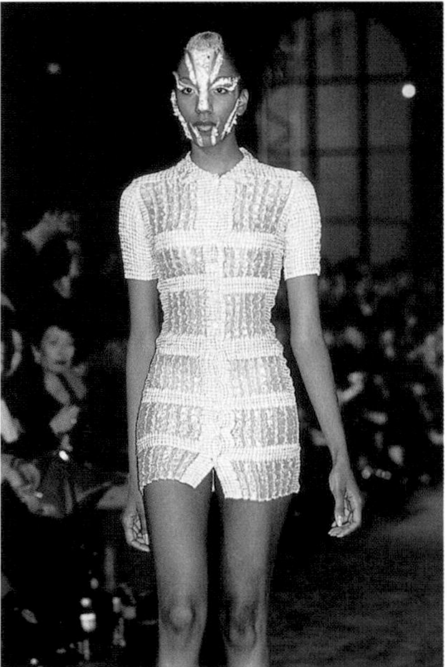

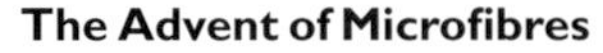

The Advent of Microfibres

On the commercial level, Japan has become a world leader in synthetic fabrics and fibre research. Ironically, this fact has its roots in the decimation of its traditional fibre industry, sericulture. At the turn of the century, silk was the mainstay of Japanese foreign trade and in 1929 '40% of all farming households in the country – roughly 2.22 million families – were raising silkworms, together producing 399,000 tons (1930 peak) of silkworm cocoons'. Immediately after the Second World War, Japan's thriving silk industry, a source of great national pride, was overwhelmed by American and European man-made fibres. The industry went into decline and in 1994 'there were around 19,000 silk farming households in Japan producing only 7,724 tons of silkworm cocoons.'[13]

The competitive pressure from synthetic fibres quickly and permanently changed Japan's silk manufacturing industry and many textile companies were forced to switch to the manufacture of silk-like synthetics. The commercial production of synthetic fibres in Japan began in 1952 when Du Pont granted a licence to Toray (formerly the Toyo Rayon Company) for the production of nylon. This was followed in 1956 by an agreement with ICI to produce polyester fibre. Today, Toray is a vitally important company in the commercial advancement of synthetic fibres.[14] Many Japanese companies excel at the technical mimicry of other natural fibres, producing convincing materials that perfectly imitate the structure, characteristics and handle of silk, fur, wool and leather. There is, it could be said, a specific Japanese sensibility which influences the designers' approach to the cloths and textures they create, taking in all the senses. Silk, for example, has an audible quality known as 'scrooping' which is considered to be especially pleasant to the ear so similar Japanese synthetic fabrics, like Toray's 'Sillook Royal', were made to imitate this aural feature; sound-wave readings were devised to verify the silk-like sound of the cloth. 'Escaine', a synthetic suede, is also a Toray fabric.

The term 'High-Tech Fibres' became generally accepted around the world by the mid-1980s, soon after the appearance of two American publications, the first a survey on 'High-Tech Fibres' and the second a book, *High Technology Fibres*.[15] By this time, many research institutes were competing to develop new 'super fibres', but Japan stole the lead in developing advanced textiles with the creation of microfibres, the key fibre-engineering breakthrough in recent years. These were not a new form of fibre but marked a new technological process for producing ultra-fine fibres that are extruded to 0.1 decitex in diameter, at least 60 times finer than a human hair. Although they are essentially our old friends polyester and nylon, if in an extremely refined guise, when microfibres are knitted or woven together they can rival all the high-touch qualities of natural materials and, most importantly, they can 'breathe'.

The 'breathability' of microfibres was a vital breakthrough in the fashion recovery of all synthetic fabrics. Finally, all the old familiar failings of stifling polyesters and moisture-trapping nylons could be forgotten. Two Japanese companies, Mitsubishi Heavy Industries and the textile maker Komatsu Seiren, announced the creation of the world's first polymer material which would function like the pores of human skin and would enable manufacturers to make 'clever clothes'. Polymers comprise long chain-like molecules that are built up from a number of repeated chemical units. When the temperature rises, gaps in this innovative polymer expand, allowing air in and water vapour out. In cooler weather, the gaps 'remember' their original shape and close to provide insulation. The Japanese company Unitika produce their own version of a skin-like fabric, 'Exceltech'. Originally used on a Chomolungma (Everest) expedition, it was made into skiwear and worn by many of the Winter Olympic medallists in 1988. It is similar to the structure of the skin but goes further, having a microporous surface that releases or absorbs water according to environmental conditions. These second-skin fabrics have totally transformed sportswear and performance clothing and 'micro' has become the fibre message of the 1990s.

The term 'shingosen' is used in Japan to denote the new tactile generation of synthetics. Kanebo is one of the leading Japanese companies and its polyester and nylon microfibre 'Bellima X' is the most popular textile product in Europe. Others of its branded fibres include the polyester-type 'Treview' and 'Vivan' and micro-fine nylons such as 'Galcem'. This new generation of synthetics have a 'natural' silk-like or 'peach-skin' handle designed specially for the fashion market.

Intelligent Fabrics and Smart Clothes

The next compelling phase of fashion's development is the reactive textile. We have come to expect more and more from our clothes, and since surprise and innovation are the very lifeblood of the fashion industry, designers are keeping a keen eye on the latest laboratory developments. Fibre technologies that can produce magical cloth is the next, and perhaps the last, great frontier for fashion novelty. Microfibres and microencapsulation can add a whole range of intrinsic functions to textiles and clothing. By shrinking the synthetic fibres to microscopic fineness, fibre technologists are making it possible to take fabrics in incredible directions, where hidden advantages can be built into fibres at the micro and molecular stage. Microencapsulation is the process by which microscopic bubbles are suspended along infinitely fine yarns and any number of chemical substances can be held within the bubbles; these chemicals are slowly released when the fabric is worn. After semi-synthetics and the first true synthetics, we have arrived at a third generation of synthetic fibres. These are no longer simply alternatives to natural fibres but are evolving as a quite new, ultramodern species of 'intelligent' textiles which have inbuilt functions. It is now possible to fill fibres with medications or vitamins that can be absorbed through the skin's pores, or to create a fabric that works in much the same way as a sun cream, allowing the tanning rays to penetrate the fibres while blocking out the harmful ultraviolet rays. Fabrics can be impregnated with fragrances, aromatherapy oils or odour banishing bactericides. The future is a potential textile utopia: temperature- or light-sensitive dyes can change colours or patterns on cloth; solar-generated heat can be stored and body temperature controlled. Textiles can enclose us in our own personalized and protective microenvironments with heat regenerative, thermochromatic, chameleonic, anti-stress, anti-odour fabrics and 'calm-inducing' or hypoallergenic fibres. Microfibres have undoubtedly improved the look and performance of synthetics, but their mystique lies in this potential to impart 'intelligence' to fibres, something new, different and intriguing – high fashion combined with high I.Q. fabrics.

The prefix 'smart' has been attached to all kinds of everyday products, from credit cards to cars, but the word itself stemmed from the materials and products developed for the military and aerospace industries and was originally used to define 'high I.Q.' weapons, buildings or constructions. As an adjective, 'smart' describes materials with integral sensors and activators, where some reactive or response function can be triggered by environmental change. The 1990s has seen an explosion of 'intelligent fibres' and 'smart fabrics', many of them emanating from Japanese companies such as Toray, Unitika and Kanebo. There is Toray's 'anti-bacteria', anti-odour, machine-washable nylon 'Dericana', and calming cloth and healing clothes are just around the corner. On a more commercial level, Toray's stain-resistant H20FF fabric is already widely available and successful. Earlier materials that resisted liquids were coated with Teflon or other finishes but this fabric uses 'water-shed architecture'. It is formed from an ultra-dense weave with millions of minute microcrimped fibre loops. These trap air which forms a natural buffer, preventing droplets of liquid from adhering to the fabric and lifting them from the surface so that they easily 'bead up' and simply roll away. Unitika has produced

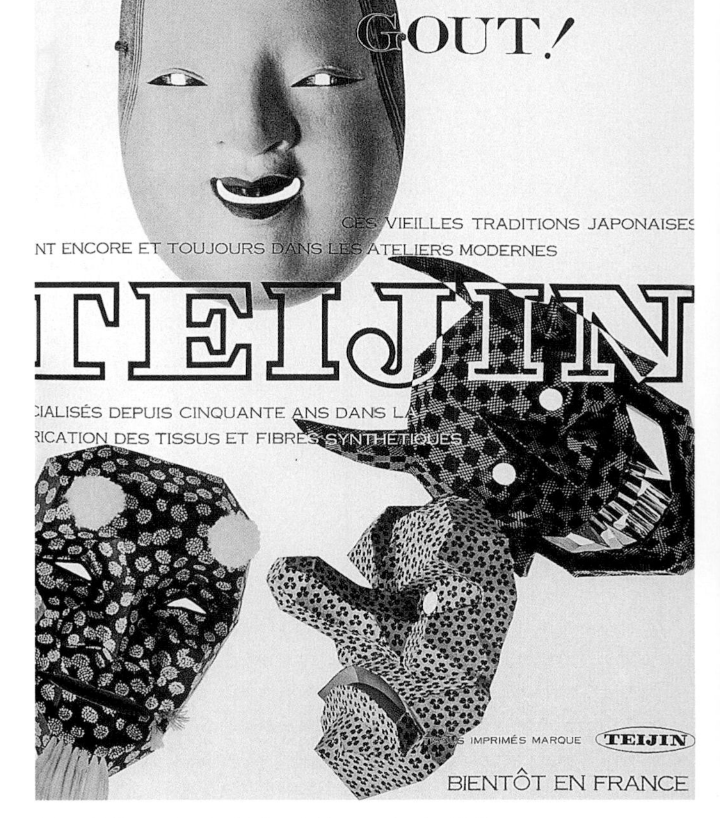

Above: Teijin as advertised in French *Vogue* curing the 1960s.
Opposite: Toray textiles promoted in French *Vogue*, mid-1960s.

'Accusol', a 'solar-powered' fabric that seals particles of zirconium carbide and other ceramic materials into fine filament fibres. This creates a cloth with a 'positive' thermal mechanism that converts sunlight into heat and stores it within the fabric to add to retained body heat, creating extra warmth.[16] In Japan, 'It has long been a dream to produce chameleonic clothes that change colour or tone according to environmental conditions such as weather or temperature.'[17] The iridescence of silk can be created by the effect of dichroism on a weave; a similar but more fundamental colour change is achieved by photochromism, where ultraviolet light changes the colour of a fabric; and thermochromism sees colour change induced by temperature or moisture.

The fashion and fragrance industries are the most complementary and intertwined of the luxury trades and many Japanese companies are actively pursuing ways of fixing scents in fabrics. Kanebo produced perfumed fibres, which they called 'Esprit des Fleurs', in 1987 and these were first demonstrated at the Exhibition of New Materials in Daily Life in Tokyo. The fabrics are 'made of fibres to which resin-made microcapsules containing perfume essence are bound. When the microcapsules are pressed and broken, the perfume is released. Microcapsule materials have been specifically developed such that the capsule thickness is accurately adjusted so that it does not break during fibre processing but will break by friction during wearing....Printing, dipping or padding is used for binding microcapsules to textiles...The perfume lasts for at least two years and can withstand several washings by hand or machine or dry cleaning.'[18] Fragrances included jasmine, rose, lavender, lily of the valley, sweet pea, sandalwood and a citrus scent. Quickly taken up by manufacturers, the market for 'Esprit des Fleurs' fibres grew rapidly and they were made into pullovers, cardigans, ties, scarves, T-shirts, handkerchiefs, hand-knitting wools, stockings, kimonos and bed linen. In 1994, the Matsui Shikiso Chemical Company described an encapsulation technique 'to develop new types of fabric that emit scents, particularly fragrances that are relaxing or that contribute to a particular ambience....the microcapsules release fragrances over a long period of time without staining the fabric base [and] can be attached to any natural or synthetic fabric without affecting its appearance or hand.'[19]

Like many technological innovations, these new fragrance-impregnated fabrics have many possible applications. They may, for example, be used in the car industry to give the interior of a modest car the smell of rich Rolls-Royce leather. In a fashion context, a number of companies are waking up to the profitable future that could result from combining a branded fragrance with a branded fashion design. Anyone who has crossed the threshold of a famous designer's salon or shop could not fail to notice an atmosphere permeated by their particular fragrance. As is common knowledge in the fashion industry, couture design is now a business that runs only at a loss and its extravagance is sustained by the licensing of products and particularly fragrances which have a huge profit margin.[20] Once a designer has gained sufficient public recognition, he or she can launch a fragrance, and from then on their economic future is all-but secure. If, as seems technologically likely, they sell clothing which is suffused with their fragrance, then the brand identity will be complete.

The only obstacle to the rise and rise of 'techno-textiles' would be a descent into inferior novelty merchandizing. There are hints of this already, such as Kanebo's 'perfumed pantyhose' or tights 'woven with microcapsules containing vitamin C and seaweed extract' (it is also hard to understand the logic behind the handkerchief 'which changes colour when wet'). At present, these products are only on sale in Japan and the Japanese are being very cautious about releasing their fibre technologies to the West – fearful, perhaps, of a boom and bust avalanche of textile gimmickry. What is clear, however, is that fabric technology has now advanced to such a point that anything seems possible and designers have before them the prospect of an ever-extending choice in man-made materials.

Design and technology in Japan are converging at a faster rate than in any of the other major manufacturing economies and 'smart-materials' research is both an industrial and a government priority. But this would make very little impact on design were it not for the fact that Japanese fashion designers themselves are so orientated towards technological and design experimentation. Apart from all the new and intriguing secret functions, synthetics have been shown to make beautiful cloth and have introduced unprecedented silhouettes to fashion design. Japanese designers have exploited synthetics to the full and have given fashion an alternative language of texture and form. Synthetics are shedding their 'poor relation' image and, since the late eighties, have been widely selected for their inherent qualities. A complete turnaround in the standing of synthetic fabrics was brought about by the impact of the work of Japan's fashion designers. No longer just the sad contents of a bargain basket in a mass-market shop, polyester and its chemical relatives are the chosen materials of the world's most successful and experimental designers. The move away from hand to mass production was an inevitable sequel to the industrial and synthetic revolutions but, in Japan, the desire to retain the mark of the hand has remained a valued symbol. It is from Japan, therefore, that the final and most ironic twist in the history of man-mades has emerged – the artificial imperfection and manufactured irregularity that signify the 'hand-made' synthetic luxury product.

Opposite: Yoshiki Hishinuma, 1998.

cheer
YES
IVORY SNOW
ERA BLEACH
YES
The Stain Lifter
The Stain Lifter
The Stain Lifter
The Stain Lifter
$2.99
5A
XTRA
XTRA
XTRA
XTRA
$4.99
$3.99

8

> 'Polyester Lives! Polyester's prices are rising – and so is its cachet. It has become a designer favorite, benefiting from fashion's passion for other synthetics such as nylon, vinyl and PVC.'
>
> *Women's Wear Daily*, May 1995

1985–
Chic Synthetic

In 1992 a *Los Angeles Times* journalist reported that 'America's most frightening fabric' was making a comeback. Polyester, the fearful fabric in question, was said to be 'fashioning a new image among top fashion designers'.[1] The fashion revival of synthetics in the nineties has been truly phenomenal and conspicuously led by many of the world's most famous and established labels including Jean Paul Gaultier, Dolce & Gabbana, Hervé Léger, Helmut Lang, and Miuccia Prada. There are many reasons for the unexpected and sweeping reappearance of synthetics in fashion, some technical, some social and others cultural. Technically, fibres and fabrics have been improved beyond all recognition since the bad old days of the seventies, but more than this, synthetics have found a completely fresh role for themselves as the materials that best communicate the complex ironic and iconoclastic messages of nineties' fashion.

The Body Perfect

The reinstatement of synthetics first began in the early 1980s and was linked to a body consciousness that was the key fashion feature of that decade. Ideal body shapes have always been socially and culturally imposed and the realization of these dreams owed much to the specialist skills of corsetmakers who, historically, were invariably men as the construction of body-shaping garments was so close to tailoring which was strictly a male profession. The perception of a 'foundation' garment changed radically during the twentieth century, going from a virtual harness for the control of wayward flesh to a near-nude body with integral fibre support. One of the key landmarks on the road to image recovery for synthetic fabrics was the mid-eighties launch of 'Lycra' – although, to be accurate, it should really be defined as a relaunch.

Opposite: Shopping Chic by Dolce & Gabbana: bustier dress in satin with platform sandals. Photographed by Steve Hiett for Italian *Vogue*, 1998.

Above: Nigel Atkinson, the inventor of synthetic illusions. Chemically sculpted velvet blouson, Italian *Marie Claire*, 1990. Opposite: Marks and Spencer Lycra hosiery promotion, 1997. Coatdress by Amanda Wakeley with Patrick Cox shoes.

Lycra is not a fabric but a synthetic elastic and one of the most promiscuous of fibres; it will happily blend and add stretch to anything, wool, silk, cotton, other synthetics and even leather, taking on the look and handle of the host material.

Initially known as 'Fiber K', Lycra was invented by Du Pont in 1958. At the time, foundation garments were made from elastic rubber which was heavy, hot and subject to degradation; in addition, rubber was prone to price fluctuations, so Du Pont's drive to create a synthetic rubber made a lot of commercial sense. Developed by the chemist Jo Shivers from petroleum-based raw materials, 'Fiber K' was unlike any other man-made material, possessing exceptional properties of stretch and recovery, stretching up to five times its original length, and it established a new generic fibre classification known as elastane or elastomeric (spandex in the United States). Girdles were considered a figure-making essential in the post-war years and the potential sales for Du Pont's new synthetic elastic were limitless. Compared to traditional materials, Lycra had considerably improved powers of what is euphemistically known as 'containment', with two to three times the restraining power of the same weight of fabric made of conventional elastics, so softer, lighter, sheerer garments could take the place of heavy, bulky girdles.

As always, Du Pont played an active role in promoting its new product to both its fibre clients and to their customers, making and presenting prototypes, carrying out extensive wash and wear tests and commissioning consumer reports, all serving to ensure that the new fibre would be a success. (Du Pont spent years on knitting tests in collaboration with Warner's, a leading company in the styling and manufacture of girdles and brassieres.[2]) Unlike rubber elastic, which was almost always covered with nylon or rayon yarns before use, Lycra could be used bare, and because it could be dyed, there was a reduced problem of 'grin-through', where the undyed elastic fibre shows through a dyed cover. By the 1960s, most bras were made of nylon or nylon lace with Lycra elastic used in the back, sides and shoulder straps. Today's ubiquitous 'body' is a descendant of Warner's Lycra 'bodystocking', first introduced in 1964. Fabric manufacturers also began to realize the potential of Lycra for swimwear, making sleek, body-fitting, lightweight swimsuits. By the late sixties the fibre was beginning to be used in outerwear and revolutionized the design of ski pants. Made in a wool/Lycra fabric, the improved slacks did not 'knee' and were extremely comfortable, becoming highly fashionable.

Tidied Up but not Trussed Up

Lycra really came into its own as fashion moved towards a more natural looking silhouette. 'Today's woman wants to be tidied up but not trussed up', said Du Pont's Marketing Manager for Intimate Apparel in 1975.[3] 'That', he continued, 'means garments that are soft, lightweight, unobtrusive, and almost nothing in weight. Modern bodygarments are not designed to eliminate the need for intelligent weight control through dietary and exercise programmes.' Fading were the days when excess flesh would be squeezed into shape with underwear; instead, taut, toned and sinuous muscles came to replace the artificial support of corsetry. A new culture of self-improvement and

Above: Claude Montana, late 1980s. Lycra's embrace, elevated from cycling shorts to haute couture.
Opposite, top to bottom:
Azzedine Alaïa, 1992. The return of the hourglass.
Helen Storey, from her collection, Edith's Sisters, 1995. A viscose/acetate hand-made lace, based on the internal structure of a rhombus cell, with viscose velvet.
Azzedine Alaïa, 1993. Fashion takes a provocative and erotic turn.
Hussein Chalayan, 1996. Bodysuit.

body maintenance spread across the Atlantic to Europe in the early 1980s. The image of the perfect muscular body became a marketing mechanism within the fashion industry and magazines strongly promoted this physical ideal. The new social and fashion pressure was to look not just thin but 'fit'. The physical control once furnished by whalebone, canvas and steel corsets or rubber girdles was replaced by the psychological self-control of diet and exercise. Increasingly during the eighties, social status came to be expressed through abstinence and being 'in shape' and the physical evidence of exertions made in the gym and the aerobics studio symbolized both leisure and expense. This was the context within which Lycra was rediscovered. What better way to show off a fashionably sleek and costly athletic silhouette than with a fabric that seemed to pour itself over the body – particularly if its fibres gave just a little shaping help. Lycra had been waiting in the wings for the aerobics and fitness decade to arrive before its second-skin qualities could be fully appreciated.

Hourglasses in Lycra

The sensuality and erotic potential of stretch fabric was introduced to high fashion by Azzedine Alaïa who launched the first of his figure-revealing collections in 1980. Dresses that seemed to be moulded onto the bodies of curvaceous models inspired the press to designate Alaïa the 'King of Cling' and conspicuously shapely women such as Tina Turner and Brigitte Nielsen were high-profile clients. However, it was not simply a question of stretching exquisite cloth over perfect bodies. Alaïa is a master tailor who does all his own cutting and prefers to work directly onto the body, in three dimensions, much in the way of the thirties couturier Madeleine Vionnet. His secret is to make the fabric move with rather than against the body and this depends on a perfect but often complex construction, with up to forty pattern pieces stitched together in a single garment. Alaïa has adapted the bias cutting and cross stitching techniques used in the construction of lingerie and corsetry to make highly fitted outerwear and many of his preferred fabrics have also migrated from the underwear drawer. Modern synthetics in the hands of modern designers have given us a new form of dress, one where the foundation garment and outer clothes have melded together.

Hervé Léger's collections of the nineties also show how the feminine body can be sculpted and held firm in a total Lycra embrace. His signature technique is to create dresses by wrapping and stitching narrow strips of bandage-like stretch rayon so that they form a flattering mould for the body. Worn by rock stars, supermodels and media celebrities, his 'bender' dresses highlight the femininity of the silhouette by emphasizing the bust and the hips, similar to the effect of the tightly laced corsets of the 1890s. The difference today is comfort, the modern factor that has come with synthetic fibres.

By the mid 1980s, fashion was drawing inspiration not just from underwear but from active sportswear, from its sleek fabrics as much as its aerodynamic forms. The skin-fit and stretch that Lycra introduced to fabrics made it the perfect fibre for its time and cycling shorts, leotards, dance, running and exercise wear became everyday fashions. Azzedine Alaïa

Above: Jean Paul Gaultier, 1989. Foundation garments become outerwear.
Opposite, top to bottom:
Azzedine Alaïa, 1991.
Dolce & Gabbana, 1990.
Dolce & Gabbana, 1992.
Dolce & Gabbana, 1992.

inspired countless chain-store stretch mini skirts and Lycra-rich fabrics brought an illusion of near nakedness to fashion. In 1986, the British designer Georgina Godley designed her first pair of Lycra leggings which were illustrated in *The Face* and *Harpers and Queen* in March 1987. 'It seemed bizarrely controversial,' she recalls, 'to put women in what were, in effect, tights.'[4] Ten years later, however, leggings were a mainstay in the wardrobes of all ages and sizes and have been one of Marks and Spencer's greatest success stories. In 1996, Marks and Spencer sold £80 million- ($120 million) worth of Lycra products, including thousands of 'bodies' – the other essential Lycra garment which Donna Karan made her signature with the launch of her label in 1985.

Lycra's success story is a marketing classic, both envied and admired by its rivals in the textile-fibre world. Du Pont have invested a fortune in a high profile marketing campaign with a strong fashion identity. The brand name has been linked with designers such as Prada, Ozbek, Gigli, Armani, Betty Jackson, Nicole Farhi, Joseph, Karl Lagerfeld, Gianni Versace, Donna Karan and Norma Kamali and their endorsements have been used as advertising material in fashion magazines. A particularly well known global advertising campaign in the mid-nineties used a series of action photographs taken by Richard Avedon. The models all wore designer garments and the cover slogan was 'Nothing Moves Like Lycra', which has been translated into many languages, including Chinese, French, German, Italian, Portuguese and Spanish. When it comes to fashion prestige, the fact that a Lycra swing ticket hangs from an Armani jacket adds reflected value to every Marks and Spencer garment with the same logo.

In high fashion terms, Lycra is synonymous with 'tight', with a photogenic sexuality, but to most people the Lycra trademark has become synonymous with 'comfort' and it seems to be (although it is not) the only spandex fibre around. There are in fact 26 producers of elastomeric fibres in the world with Lycra claiming about 30 per cent of the market.

The Babe Wave

Lycra's real success lay in its transition from a lingerie to a high-fashion fibre and the fashion cliché of the 1980s became the 'underwear as outerwear' phenomenon. Elasticated fabrics were *the* essential ingredient in putting the body back into the fashion spotlight and, more than any other designer, Jean Paul Gaultier stripped fashion back to its literal foundations. By taking what were usually hidden, intimate undergarments and representing them as outerwear, he made it clear that fashion taboos were there to be broken. Inspired by the memory of the elaborately made corsets worn by his grandmother, Gaultier brought the forms and fabrics of corsetry into a fashion context, but his blend of wit and eroticism was more than just a fashion statement. The famously provocative corset he designed for Madonna's 'Blonde Ambition' tour in 1990 somehow expressed an alternative and new vision of femininity, one that was sexually aggressive and in control.

The overt sexiness of Alaïa and Gaultier's designs reflected the altered mood of the times, when the whole issue of fashion and feminism began to undergo a profound change. Formerly, the womens'

Above:
The New RenaisCAnce couture, 1992. A celebration of England as a quaint, living museum, designed for an exhibition at Kensington Palace, London. Bodice made from pages of Shakespeare, heritage postcards and stamps, a 'Tudor half-timbered' lattice of 'Made in England' tape and dripping with plastic souvenir charms. A border of synthetic flowers represents 'The Garden of England'. Photographer: Zanna.
Detail above:
The New RenaisCAnce, 1992. 'Train of Thought', a mosaic of members of the Royal family made from laminated plastic printed squares. The collar is made from London Underground maps. Photographer: Zanna.
Opposite:
The New RenaisCAnce 1994. 'No FT, No Comment' dress, symbolizing the lifestyle of the supermodel. Skirt made from pages of the *Financial Times* and a bodice of dollar bills. Worn by Naomi Campbell. Photographer: Robert Fairer.
Overleaf:
Alexander McQueen, Devon for *Visionaire* magazine, 1996. Photograph: Nick Knight, Art Director: Alexander McQueen.

movement had rejected figure-controlling underwear and shunned artificial 'beauty aids' as denigrating, but during the early eighties a so-called 'pro-sex' feminist movement began to emerge. Later known as the 'babe wave' and promoted by writers such as Naomi Wolf and Camille Paglia, it became ideologically acceptable for liberated women to dress in a sexually suggestive way. Feminists, once regarded as the outright enemies of 'glamour', were now adopting stiletto heels, lipstick, Westwood bustiers and 'balcony' bras. Suddenly, all the old rules went into reverse. Fashion was more than an instrument of seduction, it was now an expression of sexual freedom. The new reasoning was that feminists no longer cared about male approval but were exercising power over males by dressing as sexual stereotypes. Fashion has become a weapon in the gender war; the corset is an emblematic suit of armour and the conical bra a metaphorical missile. An outbreak of academic works has emerged, mostly from American universities, exploring the semiological and sociological meanings behind Madonna's appearance. Often shrouded in the opaque language of post-structuralism, the system of analysis invented by French philosophers like Lacan, Derrida and Foucault, these writings portray Madonna as the ultimate post-feminist icon largely because of her provocative dress.[5]

Irony on the Catwalk

Reading meanings into fashion is a favourite preoccupation of our times but the important thing to remember is that the purpose of the fashion show has fundamentally changed since the 1980s and, consequently, so has the character of most couture collections. Once, the fashion show was a simple showcase, motivated by the objective of selling clothes, whereas today it is a media extravaganza: a social-status circus, a theatre for sensational outrage, and, primarily, an expensive marketing tournament dedicated to the authentication of a particular brand name. Fashion is less and less related to clothes and more and more connected to image making. Like fine art, the fashion industry is merging with the commercial media and the emphasis is on concept rather than content. Fashion shows are near cinematic total experiences and it is not surprising that many of the high-impact couture designers started their careers in the theatre or in fine art. Thierry Mugler, a former dancer, has, since the seventies, absorbed

THE NEW R SCANCE

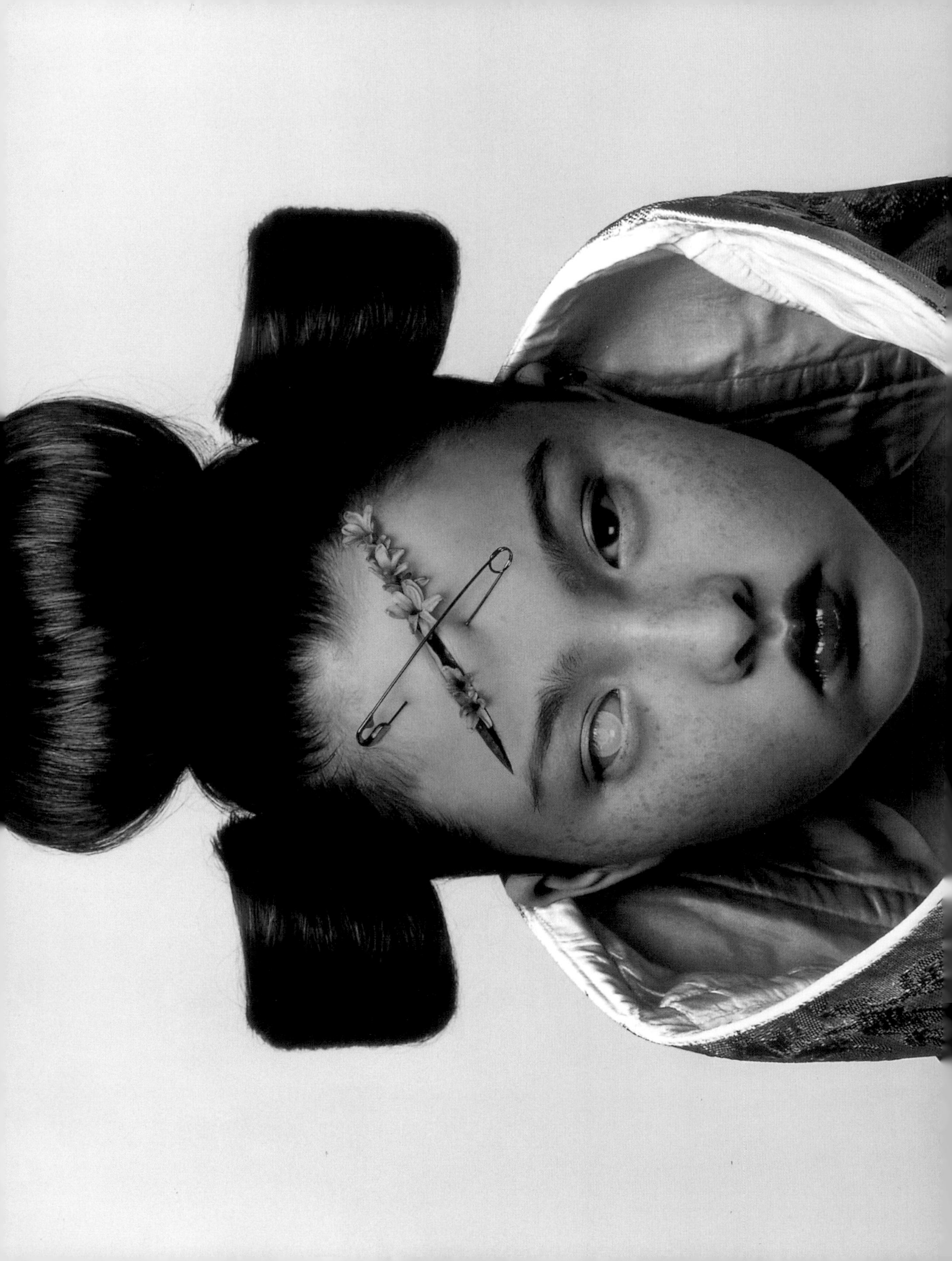

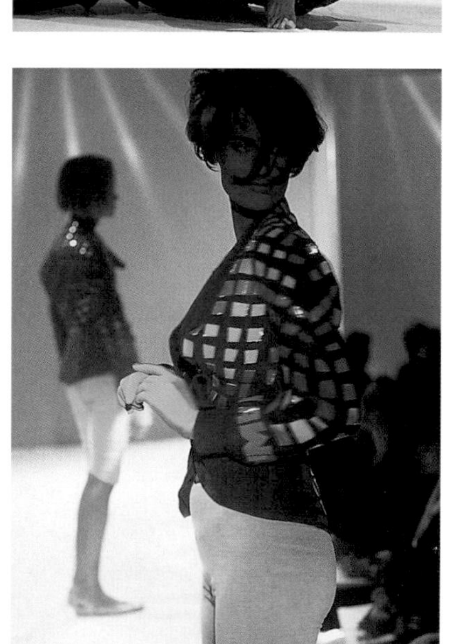

Top:
Thierry Mugler, mid-1980s. Moulded plastic bustier.
The New RenaisCAnce, 1994. A completely embroidered fabric, developed by the designers: rayon mesh combined with neon polyester and metallic threads that glow in the dark.
Photographer: Robert Fairer.
Karl Lagerfeld, 1993. Inflated crinoline.
Centre:
Moschino, 1996. Inflated PVC stole.
Moschino, 1994. Binbag Couture.
Moschino, 1995. Cheap & Chic.
Bottom:
Alexander McQueen, 1995.
Thierry Mugler, 1990s.

the drama and illusion of the theatre into his fashion design. Screen and rock goddesses, science fiction, car styling and fetishwear are just some of the sources Mugler has used to invent his catwalk cast of erotic female icons.

It was during the 1980s and 1990s that fashion really began to question its own myths and realities. What was fashion really about? Who was it for? What was it worth? Could cheap, synthetic, kitsch materials really ever merit a place within the elegant, sophisticated salons of haute couture? Franco Moschino, this century's premier fashion iconoclast, first began to challenge the elitism of the hierarchical system in 1983 when he launched Moschino Couture, followed by his diffusion line Cheap & Chic in 1988. Moschino had originally studied fine art and was to become the Marcel Duchamp of the fashion world. Plastic fabrics combined with perfect tailoring and humorous iconography were among his chosen weapons in his irreverent battle against the fashion establishment. Ironically, in the exact pattern of Duchamp, the more he parodied the sacred tenets of the elitist world of fashion, the more he was revered and admired by the very targets of his ridicule. 'Junk' materials like plastic novelties, animal-print PVC, lurid polyester and shiny nylon were transformed from sleazy to chic by being paraded in front of the world's most privileged audiences by Moschino. He made inflatable beach rings into hats and sold expensive belts made from letters spelling out 'waste of money'. One of his favourite ironic tactics was to juxtapose incongruous materials and imagery, such as mixing cheap plastic with real fur, and to elevate banal objects, using plastic fried eggs as decoration or aluminium kettles in place of Hermès handbags. He used endless visual jokes to protest against the 'fascism' of the couture system and to mock the logo addicts and label junkies of fashion.

Moschino's viewpoint was that fashion should be about fun, freedom and fantasy rather than about the display of wealth. This is an attitude which is much more acceptable in London than in the ultra-conservative salons of Milan and the work of British designers often extends far beyond fashion and spills over into other media. The New RenaisCAnce (Harvey Bertram-Brown and Carolyn Corben) are perhaps the best example of designers who have found a more universal way of communicating their ideas by moving into art direction and film making. Their design collaboration started with fashion, and they still design for the Moschino label, but their creativity has overflowed into television, media artwork, pop videos and retail design. Storytelling is very much a part of British design culture and The New RenaisCAnce fill their synthetic paradise with heroines dressed in artificial flowers and all kinds of materials generously packed with meanings and metaphors.

Jean Paul Gaultier, 1996. PVC.

The Gender of Cloth

During the nineties just about every fashion convention has been turned on its head but, again, more than any other designer, Jean Paul Gaultier has shattered the stereotypes of gender in dress. He has put men into skirts, high heels, fishnet stockings and tutus and continually switches the traditional fabrics of mens' and womenswear. Fabrics have their own erotically charged meanings and Gaultier's collections question why men shouldn't wear lace shirts, lingerie fabrics or chiffon trousers. Repeatedly he has redefined the clichés of dress, but however unorthodox his designs become, they are always perfectly tailored and wearable, a legacy of his early training with the couturiers Pierre Cardin, Jacques Esterel and Jean Patou.

Jean Paul Gaultier is technically brilliant and has always worked with the newest synthetic yarns and fabrics. The famous Swiss fabric company Jakob Schlaepfer has been providing Gaultier with textiles since the mid-1980s and their range of experimental synthetic materials is unique. Virtually all the top couture houses, including Dior, Chanel, Gigli, Ungaro, Rabanne and Lacroix, turn to Schlaepfer for synthetics with a difference: nylon that looks exactly like fur; cellophane shaped into Chanel-style pleats; huge sequins printed over with a tiger- or snakeskin pattern; plastic feathers and sequin tweeds – the possibilities are infinite. Gaultier has also used unconventional synthetics in fashion; neoprene has been made into catsuits and, like several other designers, he has borrowed fabrics from the fetishists' wardrobe. Rubber, vinyl and PVC have long-standing erotic connotations but have been socially tamed by their association with high fashion. Milan-based designers Dolce & Gabbana exploded their own variation of the Southern Italian *La Dolce Vita*-style sex bomb onto the fashion scene during the mid-eighties. Voluptuously dressed in shiny leopard-skin plastics, PVC and body-revealing stretch fabrics, their models stood in striking contrast to the demure Armani women of the time in muted wool and linen.

Gaultier has often said that London clubs and street fashions are a continuous source of inspiration for him, the places where fashion anarchy flourishes. Here he finds the raw free-spirited and eclectic ideas that become the starting point for his rebellious but glamorous collections. Many synthetic fabrics have found their way back into the fashion spotlight via street and clubwear. John Richmond's biker chic, fetish references, Destroy label, bondage chains, zips, tattoo prints, leatherette and clinging Lycra transfer the subversive symbols and edgy fabrics of youth culture and rock music to the catwalk. Like Gaultier, he questions sexual stereotypes, using ironic mixes and 'feminine' fabrics to make menswear.

Shaped by Women

In Vivienne Westwood's opinion, 'fashion is the enemy of comfort'. This was certainly true of the fashions of the eighteenth and nineteenth centuries, the periods from which Westwood has drawn much of her design inspiration, when women were virtually

Above, top: Jean Paul Gaultier, 1995. Neoprene catsuit.
Above, bottom: Jakob Schlaepfer printed plastic fabric for Paco Rabanne, 1997.
Opposite, top:
Jean Paul Gaultier, 1996. Latex.
Stephen Fuller, 1996.
Tristan Webber, 1998. Fashion borrows from the fetishists wardrobe: rubber, vinyl and PVC.
Opposite, centre:
Jakob Schlaepfer, couture synthetics for Christian Lacroix, 1997.
Pam Hogg, 1990.
Owen Gaster, 1996. Reflective and holographic synthetics.
Opposite, bottom:
Helen Storey, 1990. Hand-painted abstract on nylon/Lycra with a viscose chiffon jacket.
John Richmond's biker chic, 1998.
John Crummay, menswear, 1998.

ROCK'N'ROLL
LONDON

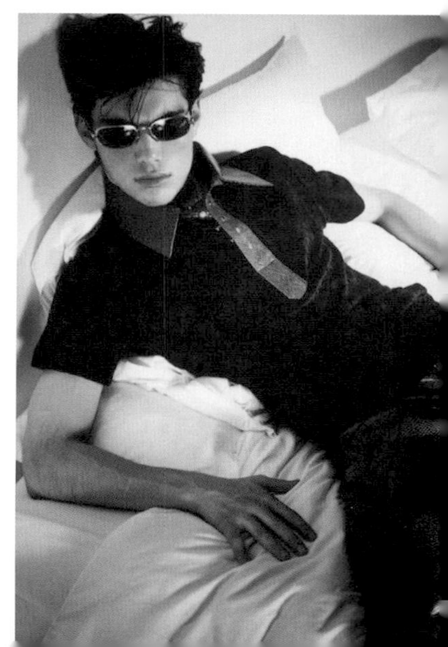

Above: Claire McCardell bathing suit, photographed by Louise Dahl-Wolfe for *Harper's Bazaar*, 1948.
Opposite, top: Donna Karan, 1991.
Opposite, bottom: Liza Bruce swimwear, 1998.

immobilized and imprisoned by their dress. But the relatively new expectation of 'fashion plus comfort' has been generated by other women designers, including Donna Karan and Liza Bruce. Both are American (although Liza Bruce has spent most of her working life in England) and their approach to fashion is one which has carried forward many of the design concepts originated by the famous New York ready-to-wear designer Claire McCardell. Stemming from a tradition of brilliantly designed sportswear and casual fashions, America's designers have introduced Europe to the idea that our clothes should be neither our prisons nor our enemies.

In the 1940s, Claire McCardell was a designer of the three-dimensional school, acutely conscious of the form of a woman's body, and her philosophy was that 'clothes ought to be useful and comfortable', like her classic jersey bathing suits and 'playclothes', designed primarily for freedom of movement. She was the forerunner of today's lifestyle designer, someone who considered the practicality of clothes within the many roles that women were expected to play. Typical was the 'Kitchen Dinner Dress' (1940) in which a woman could 'both cook and serve her guests'.[6] McCardell's jersey sheath dress became 'an American uniform' and she introduced leotards and ballet pump-style shoes into the everyday wardrobe – forty years before the aerobics era united gymwear and fashion.[7]

As a woman fashion designer in the forties and fifties, Claire McCardell was part of a very small minority. Top name designers were invariably men and perhaps the most controversial remark on the subject came in 1954 when the French couturier, Jacques Fath, told the United Press, 'Women are bad fashion designers. The only role a woman should have in fashion is wearing clothes...some day all great designers will be men'.[8] However, Valerie Steele has noted that, since 1985, the 'gender gap' is closing and the ranks of successful women designers are steadily increasing.[9]

Like McCardell, Donna Karan designs clothes that match the realities of her own, successful working-woman's lifestyle. Time has become the most precious commodity of the 1990s and increasingly emphasis is being put on 'convenience', on freezer-to-microwave ready-made meals and convenient-to-wear clothing. Designing around a total lifestyle is a modern fashion phenomenon and revitalized synthetics have contributed significantly to the career woman's wardrobe. Simplified clothes like Lycra bodysuits, stretch separates and smart separates made from forgiving synthetic fibres that improve the body's imperfections, or suits that are wrinkle-free without resembling ironing boards, have undoubtedly helped women to adapt to the complex demands of travel, work and home life.

Donna Karan began her phenomenal career in fashion as an assistant to Anne Klein, one of America's most influential sportswear designers. Her success has been linked to the rise of the 'fortysomething' female executive: 'At a time when the strict, man-tailored Dress for Success look was getting tired, and when executive women no longer felt so much pressure to look like men, Karan developed a sophisticated, sensual alternative to the business suit. Based on her own experience, Karan suspected that women would appreciate a system of dressing that was as easy as menswear, while also retaining the comfort and sensuality of clothes to fit a woman's body.'[10]

The streamlined clothes made by Liza Bruce stemmed from her sophisticated designs for swimwear. Synthetic materials, minimalist forms with precise tailoring construction and a fascination with the ergonomics of the body have all contributed to make her ready-to-wear collections the definitive examples of pared-down modernism. Gone are superfluous zips, fastenings and fussy decorative details; instead, Liza Bruce relies on her obvious expertise with fabric and selects textiles that best combine function with highly tactile and aesthetic qualities. Her unmistakable trademarks are lustrous colours, liquid-like metallic fabrics and textured Lycra-enhanced materials, made into sculpted, lean, body-caressing clothes. Often the minimalist tag is attached to designs that lack sensuality but Liza Bruce has shown that the opposite can be true.

Modernists

Better dressing through chemistry returned to the United States and Europe in the form of Japan's microfibres which migrated to the West around 1990. By then, the thinness of a single fibre was so impressive that researchers were able to drill a hole through the middle of a human hair and fit a polyester fibre inside it.[11] Microfibres are so versatile that they can be made to look and feel like anything, from delicate silk to velvety suede, and yet still retain their inherent toughness. The critical disadvantage that microfibres had to overcome was their roots, the fact that they are polyester or nylon, as the identity of fibres must be disclosed on garment labels – 'It's. . .like being named Al Capone Jr' commented John Roberts, a marketing manager with Hoechst Celanese.[12] Once again, fibre manufacturers had to put themselves in the role of 'educators' to the fashion shopper, a patently necessary expense since in one survey consumers thought that microfibre was a new breakfast cereal.[13]

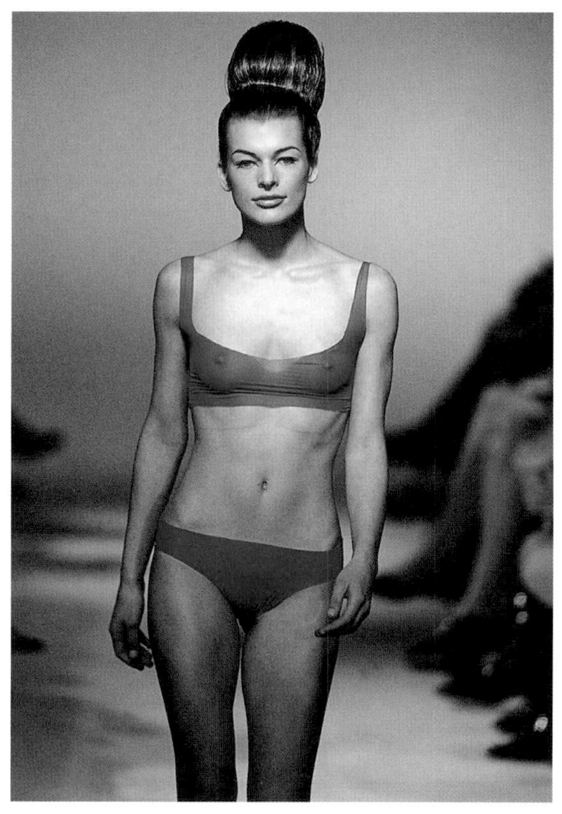

To soften the synthetic blow, the technically superior, comfortable and beautified new micro-polyesters and micro-nylons were given suitably alluring brand names such as 'Micromattique', 'Trevira Finesse' and 'Tactel'. In 1983 ICI introduced Tactel, an ultrafine variation of nylon which found its first application in performance- and sportswear before being launched in the lucrative fashion context, by Du Pont, during the 1990s.[14] 'Nylon, a fibre which has been shunned by fashion designers since the sixties, is enjoying a new lease of life', reported the *Independent* in 1993. 'Designers are queuing up to use Tactel, a new-generation nylon fibre renamed by its manufacturer, ICI Fibres, to make it more palatable to the fashion world. Among the designers using Tactel are Paul Smith, Helen Storey, Katharine Hamnett, Ally Capellino and Joseph. Tactel, say designers, could be as influential as Lycra, the stretch fibre that was the success story of the Eighties.'[15]

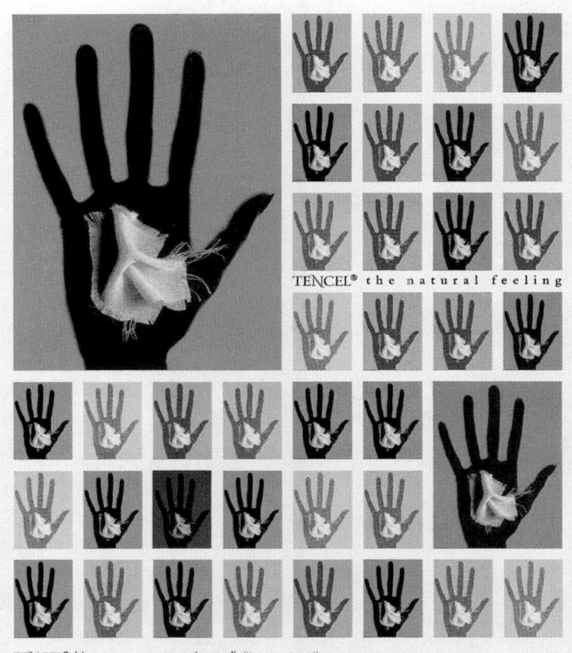

Above: Promotional campaign for Courtaulds' viscose fibre, Tencel, 1998.
Opposite, top to bottom:
Helmut Lang, 1998.
Prada, 1998.
Justin Oh, 1997, viscose polyamide dress. Simplicity arrived at through complex cutting and draping. Textile print design in light-reflective ink.
Prada, 1998.

However advanced performance fibres and fabrics may become, the one thing that guarantees their manufacturers a lucrative future is a link with glamour, and it was the re-marriage of high-technology with fashion that finally drew synthetics out of the shade. The Austrian designer Helmut Lang is one of fashion's modernists, famed for his innovative and groundbreaking use of hi-tech fabrics, mixing cheap synthetics with luxurious textiles and PVC with metallics and layers of transparent nylon. His deceptively simple cuts and minimalist elegance throws the design emphasis onto the cloth and Lang has been at the forefront of the fashion revival of synthetic fabrics. Image is everything in fashion and the concept of worth in fabric has undergone a dramatic revolution during the 1990s. Silk, once beyond the reach of ordinary people, can now be bought in supermarkets, and a silk blouse manufactured in China can cost as little as £25 ($45). Designers have the power to completely alter popular perceptions of textiles and the overwhelming power of the designer's endorsement was best illustrated when Miuccia Prada succeeded in turning nylon into 'the cashmere of the 1990s'.[16]

Miuccia Prada has also transformed what was originally a family business making luxury leather luggage into the pre-eminent fashion brand label of the nineties. Her first big success was with a black rucksack made from waterproof industrial parachute nylon and it was ownership of a Prada bag that soon became the ultimate status symbol in the fashion world. During Fashion Week in Milan the scrums outside her four Milan stores are legendary; security guards police the crowd who are set on buying one of her £400 ($650) limited edition bags. Prada clothes signal subtle good taste, understatement and discretion rather than overt sex appeal, a philosophy which probably stems from Miuccia Prada's unusual route to fashion's summit. Once an aspiring politician, she received her Ph.D in Political Science from Milan University and worked briefly for the Communist Party before she began designing for the company. Shoes were her first addition to the Prada range in 1985 and in 1989 she expanded into womenswear.

'The reason why Prada works is because it whispers, it doesn't shout', commented Miuccia Prada, 'If you want to be recognized wearing my clothes, you can be. And if you don't you don't have to be.' Prada's quiet collections have created something of a din in business circles, with profits in 1996 of £400 million ($650 million), up by 70 per cent from 1995.[17] It is the quality of Prada products that has reinvented nylon as a luxury material and this quality has a lot to do with the tradition of small scale, skilled artisanship in Italy. Downmarket Prada copies have flooded every high-street chain store in Europe, which has helped to give synthetics a 'designer' image, finally shaking off their reputation for shoddiness.

Eco-Chic

The 'greening' of textiles became a topical issue in the eighties and the most hotly debated question was whether man-mades or naturals were the most ecological fibres. It is generally assumed that natural fibres are better for the environment but the reality is that all textile production is a relatively toxic

Above: Clements Ribeiro in Tencel, 1997, as seen in British *Vogue*. *Opposite:* Deborah Milner, 1998. Dress made from ultra thin strips of 'Rigilene'© boning, a rigid plastic monofilament woven together with polyester that is traditionally used in corset manufacture. Photographer: Stuart Weston.

business.[18] Even natural fibres like cotton seriously pollute the environment during the production process as vast quantities of pesticides are used while the crop is growing and other chemicals are needed for bleaching and dyeing. Many companies and scientists are working on projects involving genetic modifications. One such agricultural scientist, Sally Fox, has developed genetically modified cotton so that it grows already coloured and Fox cottons have been used in Esprit collections in shades described as caramel, cream, coyote, buffalo and peapod green.

The old rivalry between Du Pont and Courtaulds has once again come to the surface with the launch of two branded fibres: Du Pont's nylon Tactel and Courtaulds' cellulose fibre 'Tencel'. Rayon, the first cellulose fibre, celebrated its centenary in 1994 and this was the year in which Courtaulds, the company that thrived on Rayon's success, launched Tencel. Some 14 years and around $60 million (£35 million) went into its development and Courtaulds have revolutionized the viscose-making process by developing a unique environmentally favourable process for organic solvent spinning, as a result of which Tencel is promoted as an ecologically clean fibre. Another chameleon fibre, the permutations of Tencel are endless; it has been used to reproduce velvet, corduroy, gabardine, twill, jersey, voile, lace and crepe. Press promotions have reported Tencel being used by environmentally conscious designers such as Katharine Hamnett and Helen Storey.

Fashion encourages 'wants' not 'needs' and sits in an uncomfortable partnership with ecology. How can the industry most founded on artifice and perpetual change ever really become a genuine friend of the earth? Cellulose fibres have one major advantage over all the textiles derived from oil – they are made from trees, a renewable resource. Sustainable or renewable sources of raw materials are the big areas for textile research in the future. In terms of synthetics, the big hope lies in recycling the materials themselves. Polyesters are already being re-formed from PET (polyethylene terephthalate) soft-drink bottles and Wellman's resulting 'Ecospun' is the first significant example of a fibre made from a recycled synthetic material. The latest research is exploring biologically engineered cotton with its own inbuilt herbicide and biosynthetics and biodegradable fibres are another experimental objective. Most intriguing of all is the proposal to 'grow' an already blended fibre, part cotton part polyester, the ultimate textile hybrid between nature and science.

I'll be your Mirror

Fabrics exist in the slipstream of the fashion industry; it is impossible to separate the fate of one from the whims of the other. The overwhelming fashion story of the last decade has been centred around textiles, both on the commercial level of brands such as Lycra, Tactel and Tencel and on the sensational fabric effects created by individual designers. Synthetics are essentially playful materials that lend themselves to manipulation: they can be moulded, inflated and chemically textured; surfaces can be holographic, liquid-like or metallic, deeply embossed or mirror smooth. The textile artist Nigel Atkinson, who has worked with designers such as Romeo Gigli and Azzedine Alaïa, was one of the first to demonstrate some of the remarkable fabric effects that can be created chemically. Using heat reactive dyes and

high-tech methods, he specializes in inventing textured fabrics that would take hours of laborious sewing or dyeing to create by hand. He is a textile illusionist, making cloths that seem to have come from some ancient medieval or Baroque palace but which in fact are fresh from the lab and are machine washable.

Seaweeds, daisies, raspberries, carrots, poppies, mint, roses and vine stems have all been shredded and woven with linens and silks to create new types of materials, and Hussein Chalayan has reinvented paper clothes using Du Pont's 'Tyvek' (which was first used for swimwear in 1971). A non-woven paper-based and polyethylene material, Tyvek is usually used for making non-tearable envelopes and protective clothing but it can be worn, washed and treated much like conventional fabric.[19] Chalayan's approach to fabrics is similar to that of an artist to a canvas; a canvas can be worked over, written or painted on while cloth can be burnt, baked and buried or patterned and stained with rusting iron filings – almost anything is possible. Another of fashion's chemical wizards, Stephen Fuller, experiments with solutions and compounds to arrive at fabric treatments that completely change the character of the base cloth.

Designers are dabbling with all kinds of unusual materials in their search for new silhouettes and tactile effects, sometimes experimenting with technical textiles and other materials more commonly used in sport, electronics, medicine, space, the motor industry and even agriculture. Deborah Milner is an exceptionally talented designer who uses her couture skills to invent dramatic collections from the unlikeliest materials and her designs from the late 1990s include sculpted dresses made from plastic covered wire, strips of film negative and the fine stainless steel mesh used in coffee filters. Most designers begin their work when they cut into the fabric but Milner starts from zero by fabricating the textile to fit the final shape she has visualized. A perfect example of this is a ballgown made from 'Rigilene'©, thin strips of polyester and plastic boning used in the manufacture of corsets and swimwear. Cheap, pliable and stitchable it may be, but it is a real feat of sculptural genius to turn this humdrum material into an ultra-chic dress. Fashion's dalliance with technology may come and go but ultimately it is the look that counts for designers and it is ironic that some of the most exclusive and elegant designs are now made from humble synthetics.

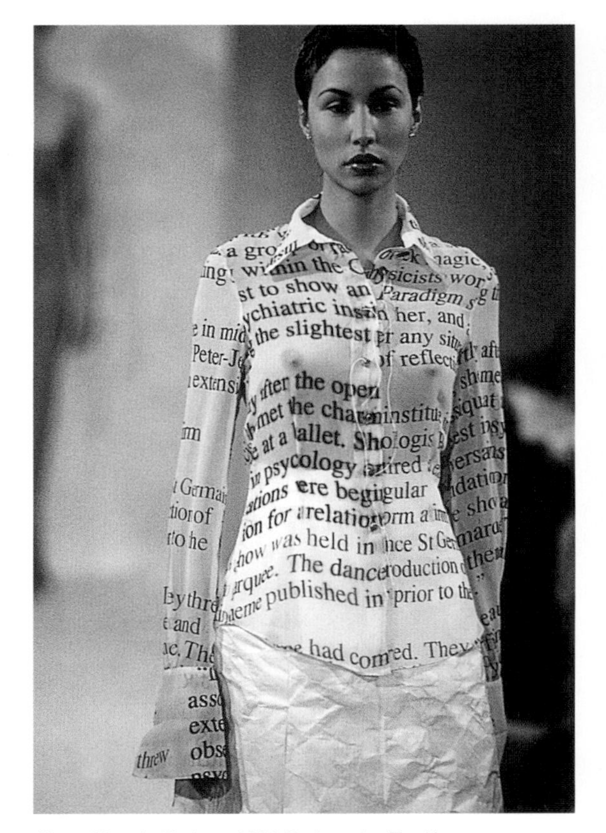

Above: Hussein Chalayan, 1994. Design using 'Tyvek'.
Below, top: Hussein Chalayan, 1995. Fabric of photographically printed paper.
Below, bottom: Nigel Atkinson, 1998. Chemically embossed textile made using heat reactive dyes.

Top:
Lainey Keogh, 1998. Hand woven dress in nylon, lurex and 'Maderia' (viscose) yarns.
Jean Paul Gaultier, 1995. Neoprene and fake fur.
Inch (Paul Helbers and Dimitry Krul), 1998. Cotton and nylon Inside Out Denim suit. Photographer: Lies Musch.
Centre:
Workers for Freedom, 1997. Rubber printed onto acrylic. Photographer: Suresh Karadia.
Workers for Freedom, 1997. Polyester and nylon. Photographer: Suresh Karadia.
Deborah Milner, 1998. Synthetic fur kimono coat. Photographer: Stuart Weston.
Bottom:
Boudicca (Zowie Broach and Brian Kirkby), 1998. 'Winged' bodice of moulded, woven plastic. Photographer: Christian Behnke.
Jo Gordon, 'Unicorn' 1998. Heat-set nylon fishing line. Photographer: Gry Garness.

9

> 'Computing corduroy, memory muslin and solar silk might be the literal fabric of tomorrow's digital dress. Instead of carrying your laptop, wear it'.
>
> Nicholas Negroponte

1990–

Science Fiction Becomes Science Fashion

A statement like this one, made by the Director of the Media Laboratory of the Massachusetts Institute of Technology (MIT) and founder of and columnist for *Wired* magazine, conjures up an almost magical vision of future textiles and clothing. Professor Negroponte is a self-confessed 'extremist' when it comes to 'predicting and initiating change', but he is only one of many futurologists and trend forecasters who see the next revolution in clothing being driven by innovations in the laboratory rather than in the design studio. If they are right, in the near future we will see the merging of clothing and technology and enjoy some sort of symbiotic relationship with our outfits. Should the digital dress become a reality, it would dwarf all the momentous advances made in textile technology in the last century.

We are, it seems, crossing the final technological frontier, where designers will be able to navigate a whimsical journey through the immaterial realities of our digitally disembodied age. Form and function are destined to become increasingly disconnected and, in the computer dominated future pictured by Professor Negroponte, our domestic objects will become silent but thinking companions. 'The wrist-mounted T.V., computer and telephone of Dick Tracy, Batman and Captain Kirk will be with us within five years and computers will be sold at fashion stores like Saks, Nike, Levi's and Banana Republic',[1] wrote Negroponte in 1996. His 'Smart House' will have the ability to sense or 'see' a human presence and anticipate its needs. Rooms will know when we eat and sleep, a 'digital butler' will vet our calls, and the telephone will not bother to ring when we are out of the house. The hundred or so microprocessors in the average home will be unified and, 'as speech becomes the dominant mode of interaction between people and machines', small consumer durables will talk and

Opposite: 'Illumination' dress, 1996, made from flexible neon fibres and polyester fabric. A collaboration between three Israeli designers, Maya Arazi, Merav Levi and Zuri Gueta.

listen to us – and to each other. The appliances in the home of tomorrow will also think for us; our refrigerators will notice that we are out of milk and page our cars to tell us to pick some up on the way home.[2] Household robots, talking toasters, mind-reading computers, televisions that can determine what we would want to watch and antenna ties – this scenario could have been scripted for an episode of *Buck Rogers*.

'If you really are curious about the future, just study the present. Because what we ordinarily see in any present is really what appears in the rearview mirror. What we ordinarily think of as present is really the past', said Marshall McLuhan in 1967.[3] Techno-enthusiasm is a recurring dream of the twentieth century and the primary periodical of the electronic generation, *Wired*, was founded in 1993 and instantly named McLuhan as its 'patron saint'. McLuhan, philosopher, media theorist, champion of aphorisms, teacher and founder of Toronto University's Centre of Culture and Technology, died in 1980, but his insights into the impact of electronic media are constantly being revisited and quoted as proverbs for the digital era. Many of the themes and speculations that are played out on the pages of *Wired* reproduce the beliefs and interpretations that also pervaded the 1960s – the last great era when youth culture and fashion became intoxicated with technology. In 1967, *The Medium is the Massage* was the textbook for an alternative, Pop-art society and *Being Digital* is today's equivalent, a sourcebook for multimedia escapists and the techno-brethren.

The Fold-up Computer

For anyone with an interest in the making of cloth and clothes, Professor Negroponte's book, *Being Digital*, unlocks the imagination; here we are offered the prospect of fashion, media and product design merging together; of fabrics becoming increasingly intelligent; and of a new hybrid species of designer, part artist, part technologist, part chemist, part computer scientist. In the future, the fashion predictors tell us, all technologies will be wearable. Digital accessories are already a part of our everyday lives, watches, personal organizers and cell phones are taken for granted and are worn or carried about in our pockets, but there is a growing belief that soon these hard-cased metal and plastic products will not only meld into one another but also melt into our

Helen Storey, from her collection Primitive Streak, 1997. The collection is based on the first 1,000 hours of human life and was devised with Dr Kate Storey, her sister, who is a developmental biologist. This piece represents cell specialization and is an embroidery of synthetic and fibreoptic fibres woven together on a hand-embroidered panel.

garments. The direct descendants of the first synthetics are today's conductive fibres which could be printed, woven or knitted into digitally responsive pockets or made into patterned and touch-sensitive electronic sections within our clothes. This is a mesmerizing thought, the stuff of fantasy and of fairy tales, of invisible, portable, thinking servants, of a personal genie in a jacket.

The 'electronic wardrobe' has only become a possibility because of the existence of synthetic materials, the microfibres, microprocessors, sensory reactors and flexible circuitboards that are the carriers and conductors of digital technologies. The realization that any technology that can be carried within microfine conductive fibres or miniaturized microchip circuitry can also be incorporated into fabrics opens up a whole new design dimension. Once again, scientist-technologists could take centre stage in the evolution of fabrics and clothing. Their contribution this time might be in the form of a microchip rather than in the chemical contents of a test tube, but it is the integration of modern gadgetry into clothes that many regard as the 'next big thing' in fashion.

In the incredible shrinking world of technology, the source of our techno-utopia is literally becoming invisible. The world's most powerful computer, which covers an area of nearly 150 square metres (180 square yards) and contains 40 billion transistors, can now be replaced by a teardrop-sized super-computer.[4] Amazing predictions are commonplace at international computer conferences and are enthusiastically reported in the press. A feature in the London *Evening Standard* in 1996 was headlined 'Pill-sized Computers: a Hard Act to Swallow' and described a minute computer that could, indeed, be swallowed.[5] Many of the expensive developments in wearable digital technologies are funded for military purposes and the computer 'pill' can keep track of a soldier's body temperature in freezing conditions. Conductive polymer fibres have been woven into long johns so that the fabric can 'radio detailed information about a wounded soldier to medics', with the fibres telling the difference between a 'high-speed round and a bayonet and if bleeding is from a vein or an artery'.[6] Such technologies will doubtless transfer, in an adapted form, to civilian life.

Lorna Ross, prototype 'glove telephones', designed at the Royal College of Art, London, 1994.

The firm belief at MIT is that the computer will become almost invisible to the user by the beginning of the twenty-first century and that passwords, keyboards and complicated interfaces will become things of the past. The next century will be the age of the ubiquitous computer, and one that is constantly shrinking. The miniaturized hybrid of the combined television, telephone and fax is in the process of being realized. British Telecommunications' Futurology Department is at the forefront of telecommunications research in Britain. Its prototype 'wearable office' is a combination of laptop, computer, digital videophone, mobile link to the Internet and a voice recognition system all squeezed into one wearable pack. 'A key difference between the "office on the arm" and the current day laptop will be the absence of a keyboard; speech recognition software will make it possible to talk to digital technologies which, it is believed, will make them more user-friendly and reduce technophobia.'[7]

For the first time, it is possible to imagine digital technologies coming in different types of packaging, in materials that may be soft and pliant, like a textile. The ancestor of all computers is the adding machine and, since the invention of the silicone chip in the 1970s, computers have evolved in the form of glorified typewriters, latterly with television attachments. Although the size and weight of personal computers has been steadily shrinking since 1980, there has been little substantial change in the way we interact with them; we are still locked into the conventional typewriter keyboard, with dimensions that have been the same for the last century.

The possibility of digital intelligence finally escaping from the prison of the plastic shell has only recently become viable. Wearable computing refers to the use of multimedia information technology with a liquid crystal display unit in a portable and personal form. During the 1990s, prototype 'radio shirts' and 'thinking boots' have been made at MIT and a 'glove telephone' was designed at the Royal College of Art in 1994. The very principles of fabric construction actually complement the absorption of information technology – fibre weaving and knitting are similar

Above: Thierry Mugler, 1991. High-fashion dabbles in sci-fi style. Opposite: Thierry Mugler, 1991. Moulded plastic bodice modelled on a printed circuitry design.

to a grid of pixels and a folding keyboard or flexible computer could be incorporated like an interfacing or lining into a garment.

A main problem and challenge of wearable technologies is how to conceal and make portable the power source for 'electronic' clothing. At the moment, the miniaturization of computer technology is not equalled by the miniaturization of power packs. The 'thinking boot' mentioned above incorporated its own energy generator – the foot in motion – and was powered by transducers built into the heels. The body's own movements generated power and Nike is collaborating with MIT to perfect the technology. If digital garments really are to evolve they too will need self-generating energy mechanisms.

Your Person-Wide Web

Most of the research into the computer/human interface is taking place in the United States and, in December 1995, Professor Neil Gershenfeld, Director of the Things That Think project at MIT's Media Laboratory, wrote an article on 'Wearable Computing' for *Wired*. Fabric, he commented, is the manufactured material to which we are most exposed, 'we wear it, stand on it, sit on it, and sleep in it, [but] all we ask fabric to do is protect us from the elements, look pretty, and not wrinkle or shrink. Can't it do more? Advances in conductive polymers and reversible optical media are pointing towards fabrics that can literally become displays. Amorphous semiconductors can be used to make solar cells to power fabric. Polymer semiconductors are candidates for wearable logic. The result would be the ultimate flexible computer architecture. Perhaps the biggest decision will be whether to buy clothes from Egghead or software from Brooks Brothers. Fashion accessories will take on new roles, becoming some of the most important Internet access points, conveniently surrounding you in a Person-Wide Web. How better to receive audio communications than through an earring or to send spoken messages through your lapel? Jewelry that is blind, deaf, and dumb just isn't earning its keep. Let's give cufflinks a job that justifies their name.'[8]

Professor Gershenfeld describes 'fibres and fabrics made from polymers and alloys which have electromagnetic properties. Fibres exist that can contract in response to electrical currents, chemicals or temperature or vary their thermal properties, and dyes that change color in response to voltage, light or heat.' He foresees a time when mini-computers might 'eventually transmit and receive data simply through human contact...or via networks hooked up to the carpet at home or in the office.... When you come home, before you take off your coat, your shoes should talk to the carpet in preparation for delivery of the day's personalized news to your glasses for reading.'[9]

Beyond shoes and clothes and well into sci-fi territory, Tom Zimmerman at MIT was reported to be working on the 'Body Bus', a digitalized human being created by using the minute nono-amp currents within the body. 'Activating your body means that everything you touch is potentially digital. A handshake becomes an exchange of digital business cards, a friendly arm on the shoulder provides helpful data, touching a doorknob verifies your identity, and picking up a phone downloads your numbers and voice signature for faithful speech recognition.... Keeping data in your body avoids the intrusion of

wires, the need for an optical path for infrared, and conventional problems such as regulation and eavesdropping.'[10]

In today's technology fixated culture, there is a belief that a better future life will come through science and technology. This century has been dazzled by its own technological progress and inspired by literary visions of the man-made Eden. Techno-utopianist writers have provided many blueprints for inventions that have actually come about and yesterday's amazing sci-fi gadgets, like dog-walking robots or 3-D videos, are today's virtual realities. Progress is no longer located in outerspace but in cyberspace and our experience of reality is an increasingly intangible one. Information is the new wealth and space is its marketplace; it is sold and transmitted beyond geographical, political or religious borders. Metaphysical space has finally been conquered and become the highway of cable television and the superhighway of information technology. Fashion is eternally mesmerized by change and is an eloquent echo of the artificial and increasingly synthetic world in which we live.

Soft Body Armour for the Twenty-first Century

A trend that exists already is the reinterpretation of fabrics intended for other industries for clothes and fashion. 'Kevlar', a Du Pont fibre that is five times stronger than steel, was 40 years in development before it was marketed in 1972. First used to reinforce Space Shuttle suits and most widely for bulletproof, stabproof vests, it has found its way into clubwear and onto the high street. Similarly, copper and even heating elements can be woven with cotton and silk to make fine cloths, while micro Phase Change Material (PCM) has its own fashion potential. PCMs are able to adapt to the environmental temperature and materially alter to create an insulating effect, turning from soft to rigid and back again as conditions dictate.

We can expect more and more from fibres and cloth: temperature changes can by indicated by colour changes, and solar energy can be stored in ceramic particles held in the fibres. Light reflective fabrics, with up to three million resin-encased microglass beads per metre (yard), are already available and interest is growing in the novelty potential of liquid crystal displays. Layers of protection could be built up and double-sided fabrics become a sandwich for all manner of functions, even air filtering systems.

Above: Vexed Generation (Joe Hunter and Adam Thorpe) parka made from high tenacity (neoballistic) nylon 66, 1994. This fabric was created by the British Ministry of Defence and is used for bomb blankets. Vexed Generation make real clothes for the twenty-first century's urbanites: dateless, seasonless, functional and protective, 'flameproof, waterproof and knife resistant'.
Opposite, top to bottom:
Yoshiki Hishinuma, 1995.
Wild & Lethal Trash (Walter Van Beirendonck), 1995. Synthetic moulded fabrics, prints of computer-generated images, accessories with built-in talking microchips, CD-ROMs and an interactive design website all bring WLT's digital fantasies home to the future-minded computer generation.
Wild & Lethal Trash, 1997. Computer culture invades the catwalk with WLT's futuristic collections.

All this is paving the way for the next phase in the evolution of textiles, towards a form of fashion where the cloth acts as a barrier against stress, pollution and bacteria as well as the ravages of the sun, rain and wind. In 1993, 'Anseline', an 'anti-stress' fabric, was launched by the French company, Avelana. The fabric contains the conductive fibre 'Resistat' which had been used for many years in technical fabrics, carpets and workwear to eliminate static shock and protect sensitive electronic equipment. The static ions given off by computer screens create magnetic fields which are believed to be harmful to people and fashion fabrics with anti-static fibres can act as a shield. The current trend is towards producing high technology and biologically engineered textiles that would act as a form of soft body armour for the twenty-first century.

The merest hint of 'smart' clothes excites the media to speculate freely on the familiar theme of technological wizardry. 'Welcome to the world of feel-good fabrics and amazing technological dream clothes', ran a British newspaper headline in 1994, reporting on the predictions made by international trend consultant Li Edelkoort. 'Our wardrobes may soon resemble medicine cabinets. Dresses made out of peat textile will protect us from bronchitis; tights will moisturise and release vitamins into our skin; cardigans will help women cope with the menopause by transmitting hormones into the body. A mad marketing ploy? Not so...we are now entering an era in which fabric will have genuine health improving qualities.'[11] Another fashion forecaster, Nelly Rodi, was quoted as saying that these smart textiles represented 'a fundamental change and within the next five years this will come through in fashion and it will be revolutionary'. Is this what we may expect the future generation of fabric producers to become – part weaver, part pharmacist and part physiotherapist?

Although the clothes shop might never replace the pharmacy, the big question for designers is how 'smart-fibre' technology might influence and change the vast global fashion system. At present, the fashion clock is set to last for one season only, but there is every possibility that this dictate could be revolutionized by textiles that can be worn across the seasons. Traditionally, we are accustomed to buying thick wool knits or weaves in the winter and fine fabrics like linens and cottons in the summer, but thermally controlled fabrics could, by keeping our

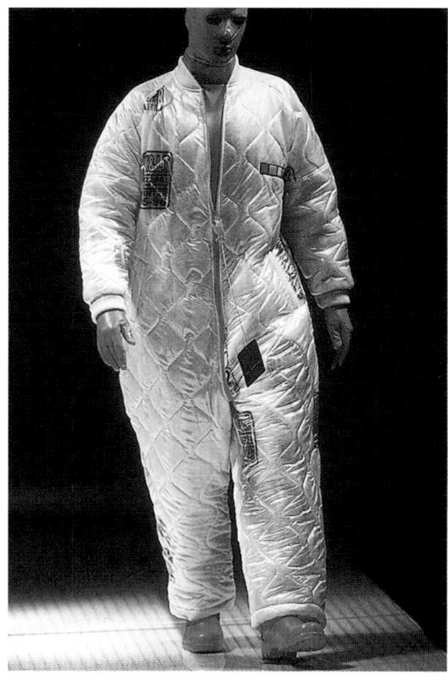

Above: Wild & Lethal Trash, 1997.
Opposite: Wild & Lethal Trash, 1998.

body temperatures constant, make seasonal collections a thing of the past. Clothes that can adjust to rapid temperature changes would mean that the same suit could be worn on a hot day in Hong Kong or a cold afternoon in Moscow. By using fabrics that retain solar energy, warmth can be stored in a fine cloth, and even clothes with their own inbuilt 'air conditioning' system could conceivably appear. If 'smart' materials really do become commonplace in fabrics, with them will come a new breed of scientist-fibre engineer to add to the design equation, someone who can literally tailor the new technologies to fit in with the old industries of fashion and textiles.

An Unlikely Marriage

The ubiquitous and invisible computer may well be a certainty but whether or not the body, and particularly clothing, will be the platform for these technologies is still open to question. Many fundamental obstacles need to be overcome before the digitalized invasion of the wardrobe can realistically take place, one of the main problems being that computer technologists and fashion designers inhabit totally different cultures. The world of digital technology is a strikingly male and science-dominated one, where the 'smart clothes' and 'wearable technologies' so far produced have not actually been clothes at all but virtual reality-style headwear and something similar to high-tech tool belts: wearable computers have simply been turned into products that are already familiar within the masculine context.

The prefix 'wearable' suggests 'soft', 'flexible', 'fabric', 'clothes' and perhaps 'accessories' to fashion designers, whereas technologists think 'function' and then encase it in hardware. However bizarre we may think MIT's objectives are in relation to integrating technologies into our everyday lives, more than 100 multinational companies have invested millions of dollars in the Things That Think project.[12] Advances in computer science are profoundly changing the artefacts that surround us and if wearable technologies have any future at all it is essential that they evolve as multi-disciplinary projects. Their viability and chances of success remain in doubt unless there is more direct contact between the makers of clothes and the inventors of technology.

Part of the current fashion intoxication with the incongruous world of high technology is that it offers the only real potential for anything radically new in clothes. The twentieth century has witnessed the most extreme changes in fashion ever known in history. It is hard to imagine a cut, a silhouette or a fabric that we do not already know or have not seen reincarnated many times over, so that the only place to go that is not a revisit is towards some form of futuristic, multi-functional fashion. Technology is the chosen route for fashion's adventurers, those with an idealistic vision of previously unimagined textiles and clothing, and only time will tell whether or not they are inhabiting the land of make-believe. Ours has become a culture located in cyberspace and the concept of the shirt-computer is very much in the spirit of the intergalactic fashions of the 1960s. Without doubt, this is an exciting time for fashion design and no one yet knows whether the much talked of 'wearable technologies' will become the future's most desirable items or just another burnt-out fad. It might be wise to remember the age-old questions put to every would-be fashion designer: who will wear it, where will they wear it and will it wash? Might we one day be putting our computers into our washing machines?

Epilogue

Ask anyone if they prefer things 'natural' or 'synthetic' and they will surely come down on the side of nature. This applies as much to the fibres they wear as to the food they eat, but words can be strangely misleading and nowhere more so than in defining today's textiles. What is natural and what unnatural in fibres, and which of our senses can truly make the distinction: sight, touch, smell? Few people can describe the difference between a cellulose and a polymer fibre and even fewer can make sense of the hundreds of different brand names given to the few generic fibres: viscose, polyamide, polyester, elastane, acrylic, acetate or polypropylene. Mystery adds attraction to materials and most people would be surprised to discover that the exciting 'new' microfibres found in today's clothes are actually just refined versions of our old friends nylon and polyester. In the Lewis Carroll-like world of synthetics, things are never quite what they seem.

Synthetics have long been dogged by the description 'unnatural' and all too often are cast in the role of nature's decoy, as the cheap mockery of an authentic product. For decades, marketing wars raged between the natural-fibre producers and the manufacturers of synthetics and closeness to 'nature' was the most potent of all the feel-good psychological weapons. 'Semi-synthetics' were believed to be marginally better than full synthetics because they are made from organic materials, but 'full' synthetics were chemistry incarnate – materials conjured up in petrochemical plants from mysterious, suspect substances. Ironically, the raw materials of synthetics are no more 'unnatural' in their origins than any other fibre. Nylon and its chemical acquaintances polyester and acrylic are made from the coal, oil and gases that are the natural derivatives of ancient organic life – from trees and plants that are just considerably older than the raw materials of 'natural' fibres and chemically deconstructed by nature itself. Technology has made the twentieth century the age of the facsimile, but, from the beginning, man-made fibres were tainted by the designation 'artificial' and it is our subjective resistance to this concept that has shaped the history of synthetics.

Opposite: The Man of the Next Century as predicted by the product designer Gilbert Rhode, photographed by Anton Bruehi and illustrated in British *Vogue* in 1939. 'His hat will be an antenna snatching radio out of the ether. His socks – disposable. His suit minus tie, collar, buttons. His belt will hold all his pockets ever did.'

Virtually everything has had an impact upon the evolution of synthetic fashion: from world economics to wars, from geological resources to youth culture; from space technology to haute couture; and from mass retailing to simple taste swings. It is the long and complex route from patented fibre to designer garment that has made the history of synthetics a complicated jigsaw to construct. Of the many forces that resulted in the commercialization of chemical fibres, it was the political tension between Japan and the United States that really brought nylon into being. Japan's vital silk exports were threatened by man-mades from the 1940s which stimulated Japanese manufacturers to develop much improved versions of synthetics, and by the 1980s, Japan had become a world leader in the development of man-made fibres.

Nylon is the grandmother of all synthetics and there can be no doubt that it was an epoch-making invention, as were those two other fibre *grandes dames*, polyester and acrylic. Here were materials without any tradition or cultural meaning and to begin with they were welcomed as liberating wonder fabrics. In the 1950s, science was the gateway to the housewives' paradise, a drip-dry, damp-cloth Eden. The children of this plastic garden created their own transient synthetic world of space and pop fashions during the uninhibited sixties. Within just a few years, however, oil spills and ecological disasters had seriously damaged the reputation of all things man-made while the oil crisis of 1973 undermined the comfortable economic stability of raw material prices. Soon, synthetics sunk to the all-time design low of the much ridiculed shell suit and to the textile anonymity of the blend. 'Back to nature' hippydom eventually evolved into a mainstream retail nostalgia for a Laura Ashley and Ralph Lauren wholesome lifestyle, filled with the 'integrity' of wool, linen and cotton. A century ago, the first man-made fibres were the standardized chemist's copy of the worm's silken thread, a cheaper, substitute material – and the failing of succeeding generations of synthetics has been their relentless uniform perfection. The fibres were too predictable for the seventies' dream of the idealized homespun lifestyle; nature was wayward, irregular and best expressed through organic fibres. It was a gloomy time for the synthetics industry and the fashion exile of man-mades was to last for more than a decade.

Not so long ago, 'natural' fibres were commonly believed to be environmentally superior to man-mades but, as the eco-investigators began to uncover the extent to which chemicals were used in the production of these 'naturals', they too were found wanting. The comparative environmental consequences between the production of natural and synthetic materials continues to be hotly debated. An assessment depends upon a complex scale of environmental damage, linked not just to the origins of the fibres and whether they are derived from non-renewable resources but also to the chemical processes used in bleaching, dyeing and finishing natural fibres (cotton, for example, requires the use of enormous quantities of pesticides to protect its growth). It is so difficult to secure hard factual information on the 'cradle to the grave' production methods of most fabrics, that the web of comparative ecological soundness has yet to be untangled.

The motivation behind much of the scientific research into synthetic-fibre alternatives was to find a way of defeating the price fluctuations connected with natural fibres which are affected by seasonal availability, disease, weather, trade and wars. At first it seemed that the chemical industries had solved national, political and economic headaches until the oil crisis of the early 1970s reminded them that they too were dependent on imported raw materials that could just as easily 'dry up', causing the price of fibres to soar. The truth is that the world is dependent on both synthetic and natural fibres and its clothing needs could not be met if we relied on one or the other.

In 1940, Du Pont asked the 18,000 visitors to its Wonder World of Chemistry exhibition, 'What do you believe is the most important future development chemistry can make for the welfare of mankind?'. Among the hopes and aspirations of America were 'dishes that require no washing, pocket air conditioners, chip-proof nail polish, plastic houses, synthetic furs, transparent steel, shoes that never need repairing, synthetic rain, artificial water, noiseless explosives, cosmetics to keep faces young, and better fabrics and longer-lasting materials'.[1] Today's prophets of the world of tomorrow predict that intelligence will soon be embedded into sensory fabrics and clothes; that fine cloth of the future will function like a personalized protective suit of armour, monitoring health or temperature and making appropriate chemical adjustments within its microencapsulated fibres. The reactive properties of natural organisms

'Paradise' dress and nature imitated: a synthetically clad Eve in an artificial Eden. Designed by Maya Arazi, Merav Levi and Ayelet Ofer Shai, 1995.

have already been emulated synthetically and woven into cloth. With the spiralling miniaturization of technology it seems perfectly reasonable to imagine a time when digital technologies might break out of their traditional plastic prisons and be absorbed into our clothing. All this is a long way from the copy-cat beginnings of art silk and the discardable fashions of the sixties. Synthetics are at last finding their natural role by making it at least thinkable that digital technologies could find their way into fabrics and ultimately clothes. Just as nylon stockings humanized and feminized the first synthetic fibre, so the predominantly masculine hard-plastic products of the digital world might well be revolutionized by the imaginative, whimsical touch of textile weavers and fashion designers.

The real danger with any new technology, however, whether fibre or digital, is that it will be devalued by being made into too many badly designed and mass-produced novelties. And, indeed, the sickness of the twentieth century is the deadening uniformity of things, the spreading homogeneity and the shrinking choice that has accompanied global mass production. Plastics and synthetics are blighted by a debased image because of the sameness of the products they create, filling our world with identical things that induce an epidemic of boredom. The availability and abundance of man-mades were both the key to their success and their downfall. In fashion terms, the eternal problem with synthetic fibres is their association with repetitive mass-produced clothing; by extension, perfection is the flaw in synthetic fabrics, the mark of their 'artificial' origins. However, in our technologically advanced age, even flawlessness can be corrected by computer programmes. The random irregularities and characteristics found in hand woven cloth can be coded into high-tech weave programmes which will create the illusion of a traditional homespun fabric from the most sophisticated of technical fibres.

Despite their arriviste status in the fashion scene, synthetics have generated more excitement and more disillusionment than any other materials previously known. Their story is infinitely complex and riddled with metaphor and paradox; of utopian inventiveness turned sour, consumer dreams turned to polluted nightmares and of a new class warfare in clothing, when the imagined democracy of luxury for all turned into polyester for the underprivileged. At the heart of the rise, fall and rise of synthetics in fashion lies their own amorphous character and the perpetual fashion swings between nature and man-mades.

There is an important social distinction between the fake textile, intended to deceive, and the openly artificial textile. The Japanese fabric designer Hiroshi Matsushita wrote that it is a mistake to use man-made fibres to simulate natural ones because 'man-made fibres should be appreciated for their distinct qualities, not for their ability to force an artificial reproduction of something natural. It is also a misconception that man-made means machine-made while natural is associated with hand-made processes. For many of our synthetic fabrics we actually use hand-made processes. With that in mind, when we use man-made textiles we do not, for instance, use a heat treatment process to make [them] "more beautiful" or to make them appear "more natural", or [to create a less expensive] substitute for their natural counterparts. Rather, we prefer to leave the synthetic textiles in their "natural" state and that is their source of beauty, that is what makes them special and appealing.'[2]

Notes

Sums of money are given in £ sterling or in US$, whichever was the currency quoted in the original source. Approximate equivalents in the second currency are only given in chapters that cover more recent decades.

1
The Chemist Conquers the Worm

1 'Lavoisier – An Estimate by Lammot du Pont, Chairman of the Board, E. I. Du Pont De Nemours & Co. (Inc.)', *Du Pont Magazine*, vol. 38, no. 3, June/July/August, 1944, pp. 2–7

2 *Du Pont, The Autobiography of an American Enterprise*, no author given, E.I. Du Pont De Nemours & Co. (Inc.), 1952, p. 8

3 For a detailed study of the du Pont early family history see *The History of the E.I. du Pont de Nemours Powder Company*, no author given, originally published by Business America, New York, 1912, reprinted by Lindsay Publications Inc., Manteno, 1990

4 *The History of the E.I. du Pont de Nemours Powder Company*, no author given, Lindsay Publications Inc., Manteno, 1990, p. 4

5 ibid. p. 32

6 The word 'dynamite' was derived from *dynamis*, the Greek word for 'power'.

7 *Du Pont Magazine*,vol. 1, no. 1, 1913

8 ibid. pp. 114–18

9 op cit. *Du Pont, The Autobiography of an American Enterprise*, pp: 71–3

10 ibid. p. 76

11 Coleman, D. C., *Courtaulds: An Economic and Social History*, vol. 2, Clarendon Press, London, 1969, pp. 2–3

12 Corbman, B.P., *Textiles: Fibre to Fabric*, McGraw-Hill Book Company, New York, 1975, 5th edition, p. 333

13 Feltwell, J., *The Story of Silk*, Alan Sutton, Gloucestershire, 1990, p. 8

14 ibid. p. 9. This great theft took place, according to Dr Feltwell, in AD 552, when two Persian monks secreted the eggs of the silkmoths in hollow canes and smuggled them out to Constantinople, at the risk of being beheaded. Persia subsequently became a great silk producer.

15 Cook, J. Gordon, *Handbook of Textile Fibres, Man-Made Fibres*, Merrow Publishing Co. Ltd, Watford,1968 (first published 1959), p. 5

16 Fraser, G. L., *Textiles By Britain*, George Allen and Unwin, Ltd, 1948, p. 93

17 op. cit. Coleman, p. 5

18 op. cit. Fraser

19 Their technique of spinning the substance into lustrous filaments sounds similar to the method for making candyfloss and it became yet another form of 'artificial' silk.

20 The chemistry of these early fibres was often complex and both of the cellulose acetates, 'Acetate' and 'Triacetate', not only differed from rayon but also from each other. They are not completely cellulose products and have their own specific properties. Unlike viscose, the filaments are not reconverted to cellulose and the extruded fibres do not enter a coagulating bath but solidify in a stream of warm air.

21 *The Origin and Development of Courtaulds Between 1700 and 1986*, no author given, Courtaulds publication, p. 9

22 Although 'Cupro' never really achieved large scale production, Bemberg later extended production to their American branch of the Bemberg Corporation in 1926 and to Doncaster in England in 1931. The primary use of its yarns was for ladies' stockings and underwear and a proportion was also used for crepes de Chine and crepes georgette.

23 op. cit. *The Origin and Development of Courtaulds Between 1700 and 1986*

24 op. cit. Coleman, p. 37

25 op. cit. *The Origin and Development of Courtaulds Between 1700 and 1986*, pp. 11–12

26 ibid. p. 10

27 Samuel Courtauld IV was a great patron of the arts, of painting, music, opera and literature and it was his art collection, along with a generous endowment, that established the Courtauld Institute of Art in London. As an industrialist, he became active during the early years of the Second World War in talking and writing about the future of industry and about the relationship between industry and government and between employer and employees.

28 'Du Pont Advertising – It's Value to the Trade', no author given, *Du Pont Magazine*, March 1919, pp. 17–27

29 *Du Pont Magazine*, January 1918, p. 22

30 Fenichell, Stephen, *Plastic, The Making of a Synthetic Century*, Harper Business, HarperCollins, New York, 1996, p. 129

31 ibid. p. 129–30

32 op. cit. *The Origin and Development of Courtaulds Between 1700 and 1986*, p. 14

33 ibid.

34 op. cit. Coleman, p. 271

35 op. cit. *Du Pont, The Autobiography of an American Enterprise*, p. 91, 'A Woman's World?'

36 'Rayon Advertising and Promotion', a report for Du Pont by the advertising company Frank Seaman Incorporated, New York, 21 October 1927, p. 1

37 White, Palmer, *Elsa Schiaparelli, Empress of Paris Fashion*, Aurum Press, London, 1986

38 op cit. 'Du Pont Advertising – It's Value to the Trade', p. 18. The article gives an impression of the scale of company publicity undertaken: 'We publish the most pretentious house organ produced, the *Du Pont Magazine*, with a monthly circulation of 250,000, the postage for which costs $5,000 per month. We also publish the *Du Pont Export Magazine*, circulating 15,000 copies per month in foreign countries. We annually print and supply through the trade tons of printed booklets, in variety as numerous as our line of more than a thousand different products demand.'

39 op. cit. Fenichell, p. 119

2
'N' Day: The Dawn of Nylon

1 'Castor Oil and Coal: Newest "Silkworms" for Stockings', *Science News Letter*, 1 October 1938, p. 211

2 Fenichell, Stephen, *Plastic, The Making of a Synthetic Century*, Harper Business, HarperCollins, New York, 1996, p. 159

3 'Nylon', *Fortune*, no author given, July 1940, pp. 57, 60 & 114

4 Meikle, Jeffrey, *American Plastic, A Cultural History*, Rutgers University Press, New Jersey, 1995, p. 133

5 *Science News Letter*, 1 October 1938

6 Wharton, Don, 'Nylon – A Triumph of Research', *Textile World*, October 1938

7 Labovsky, Joseph, 'A Short Biography of Nylon', no date, p. 2. Labovsky was a scientist who worked on nylon research.

8 ibid.

9 Albee, George, 'Nylon's Success Story', *Du Pont Magazine*, vol. 38, September/October 1944, pp. 8–10

10 op. cit. Meikle

11 ibid. p. 134

12 ibid. p. 133

13 ibid. pp. 133–4

14 '$10,000,000 Plant to Make Synthetic Yarn; Major Blow to Japan's Silk Trade Seen', *New York Times*, 21 October 1938

15 *Du Pont, The Autobiography of an American Enterprise*, no author given, E.I. Du Pont

De Nemours & Co. (Inc.), 1952, p. 92

16 op. cit. '$10,000,000 Plant to Make Synthetic Yarn; Major Blow to Japan's Silk Trade Seen', *New York Times*. This report predicted a major upheaval in the economic position of Japan as the key American manufacturers prepared to produce nylon: 'Celanese Corporation of America, manufacturer of acetate-process yarns and fabrics, announced plans for erecting near Pearlsburg, Virginia, a $10,000,000 plant for production of an entirely new synthetic yarn that can be used in all textile fields. The Celanese announcement followed closely a similar statement by E. I. du Pont de Nemours and Co., of plans for building a $7,000,000 plant near Seaford, Delaware, for production of textile yarn reported to be its now-famous "Yarn 66", a new synthetic fiber also adaptable to various textile uses but intended chiefly for hosiery.'

17 op, cit. Fenichell, p. 139

18 Clapper, Raymond, 'Artificial Silk Worm Developed by Du Pont', *New York World Telegram*, 17 January 1939

19 op. cit. Wharton

20 op. cit. *Du Pont, The Autobiography of an American Enterprise*, p. 106

21 Coleman, D. C., *Courtaulds: An Economic and Social History*, vol. 2, Clarendon Press, London, 1969, p. 385

22 op. cit. 'Nylon', *Fortune*

23 *25th Anniversary of Du Pont Nylon, 1939–1964*, press release, October 1963

24 op. cit. Fenichell, p. 140

25 ibid. pp. 141–2

26 op. cit. Meikle, p. 138

27 ibid.

28 Letter held at the Hagley Museum and Library between ICI and the Du Pont Foreign Relations Department from Mr P. H. Chase, dated 17 September 1941

29 Letter in the Du Pont company archives at the Hagley Museum and Library.

30 op. cit. Meikle, p. 137

31 Dutton, William S., 'New York Presents "A Drama of Opportunity"', *Du Pont Magazine*, vol. 33, no. 6, June 1939, pp. 1–8

32 ibid. 'Lucite' was described as a 'methyl methacrylate plastic, offspring of the coal-air-water trinity...[It] is several years old but the uses to which it is being put are as new as nylon. "Lucite" has unusual optical properties and will convey light around curves and bends. It is light in weight, substantially unbreakable, water-white and as transparent as crystal.'

33 ibid. p. 2

34 ibid.

35 Simpich, Frederick & Culver, Willard R., 'Chemists Make A New World' and 'From Nature's Building Blocks', *National Geographic*, vol. 76, no. 5, November 1939, pp. 601–40

36 op. cit. 'Nylon', *Fortune*

37 op. cit. Simpich, pp. 631–2. Du Pont had also 'made a sponge that never saw the sea bottom: yet it would fool even a fish, it looks so natural. It is from spruce. It can be cleaned and sterilised by boiling and won't crumble. Yet it will absorb twenty times its own weight of water.'

38 op. cit. *25th Anniversary of Du Pont Nylon, 1939–1964*

39 ibid.

40 ibid.

41 ibid.

42 op. cit. Fenichell, p. 145

43 op.cit., Simpich, p. 624

44 *Business Week*, 18 January 1941

45 ibid.

46 Du Pont spent $11,000.000 on nylon plants, one in West Virginia, which partially processed the raw materials, and one in Delaware, which turned out the finished nylon. They employed 830 men and turned out approx. 2,070,000 kg (4,600,000 lb) of nylon a year.

47 op. cit. 'Nylon', *Fortune*, p. 57

48 Press release issued to magazine and newspaper editors by the Du Pont Public Relations Department, 9 May 1940

49 *Department Store Economist*, 25 May 1940, p. 11

50 ibid. p. 8

51 op. cit. Meikle, p. 147

52 op. cit. Albee, p. 8. The scale of nylon's success is summed up: 'Before long, nylon had created a full-scale industry employing more than 3,500 men and women directly and creating jobs for thousands of others in hosiery and textile mills. Between January of 1940 and February of 1942, it is estimated that 14,000,000 dozen pairs of nylon stockings were sold in the United States and 80 per cent of the better toothbrushes were bristled with nylon, and 50 per cent of all the hairbrushes. Manufacturers of lingerie, neckwear, bolting cloth, rainwear, screen printing cloth, sewing thread and many other products formerly employing silk recognized nylon as a superior fiber for their purposes.'

53 ibid. 'The Notre Dame eleven wore them in its game with Army at Yankee Stadium a little more than a month before Pearl Harbor. It was a scoreless clash played in teeming rain and clinging mud. The nylon pants, which weighted only four and a half ounces dry compared to 11 or 16 ounces for ordinary moleskins, proved a boon in the quagmire because they retained little moisture and picked up little mud. With the onset of war and the allocation of nylon for more vital needs, the pants took on added value. The same ones were worn by the South Bend varsity for six seasons.'

54 ibid. p. 9

55 *Du Pont Magazine*, vol. 36, August/September/October 1942

56 op. cit. *25th Anniversary of Du Pont Nylon, 1939–1964*

57 ibid.

58 *Du Pont Magazine*, vol. 40, December 1946, pp. 15–17

3
Better Living through Chemistry

1 *Du Pont, The Autobiography of an American Enterprise*, no author given, E.I. Du Pont De Nemours & Co. (Inc.), 1952, p. 91, written in 1947 by John Gunther, published in *Inside USA*

2 Galbraith, J.K., *The Affluent Society*, Penguin, London, 1958

3 Dutton, William S., 'Tough jobs are their dish', *Du Pont Magazine*, April/May 1952, p. 3

4 Statistics show that by 1972 polyester had overtaken nylon in world production terms and that by 1992 it accounted for 75 per cent of all synthetic fibre production. For further statistical information, see F. T. Brunnschweiler, David & Hearle, John, 'The Polyester Story', *Textile Horizons*, June 1992, p. 22

5 Polyester fibres and polyethylene film are both made from ethylene glycol – essentially, motorists' antifreeze – added to terephalic acid, both of which are derived from oil. The two petrochemicals are heated in a vacuum and the polyester polymer is formed by a reaction between them, which results in polyethylene terephthalate.

6 'The World of Silk,' *American Fabrics*, no author given, vol. 15, Fall 1950

7 ibid.

8 *Du Pont Magazine*, October/November 1951, pp. 8–9

9 *Du Pont Magazine*, April/May 1954

10 'Wardrobe Revolution', *Better Living*, May/June 1959

11 Rothgerber, Jr, Leonard A., 'Travel Light...And Love It', *Du Pont Magazine*, August/September 1956

12 *Du Pont Magazine*, vol. 49, August/September 1955

13 'How You Help to Tailor Textiles', *Du Pont Magazine*, April/May 1958, p. 12

14 Heine, Emily, 'Wash-and-wear Goes Automatic', *Du Pont Magazine*, pp. 2–5

15 'Wash-and-wear Suits', *Du Pont Magazine*, April/May 1953, pp. 32–3. A similar Witty suit was worn on Broadway by a performer in the musical, *Wish you Were Here*. During each performance the actor was dropped, fully clothed, into a swimming pool. The suit dried out between shows and the actor wore it repeatedly without pressing.

16 De Llossa, Martha, 'The Story of Polyester', *American Fabrics and Fashions*, no. 132, 1985, p. 14

17 'Wardrobe Revolution', *Better Living*, May–June 1959, p. 16

18 ibid. p. 13

19 *Du Pont Magazine*, vol. 35, no. 5, 1941, p. 7

20 Meikle, Jeffrey, *American Plastic, A Cultural History*, Rutgers University Press, New Jersey, 1995, p. 186

21 op. cit. De Llossa

22 op. cit. Meikle, p. 185

23 *Better Living*, September/October 1959, p. 17

24 ibid. p. 16

25 ibid. pp. 16–19

26 op. cit. 'How You Help to Tailor Textiles', *Du Pont Magazine*, p. 13

27 ibid. p. 11–13

28 'New Trends in Blended Fabrics', *American Fabrics*, no author given, vol. 20, Winter 1951–2, p. 129

29 'American Fabrics Presents the Key to Today's New Combinations of Three Great Fibers', *American Fabrics*, no author given, vol. 20, Winter 1951–2, pp. 43–50

30 'The Sweatered Look in Orlon', *Better Living*, January/February 1959, pp. 12–13

31 *Du Pont Magazine*, vol. 54, no. 2, March/April 1960

32 *The Ambassador*, no. 3, 1953, pp. 102–3

33 *Better Living*, January/February 1959, p. 15

34 ibid.

35 Interview with Peter King, London, November 1995

36 Ashton, Lady, '"Terylene" – what's in it for you?', *Queen*, 28 July 1954, p. 45

37 Ashton, Lady, 'Curtain up on "Terylene"', *Queen*, 22 September 1954, p. 10

38 Dutton, William S., 'Tough Jobs are their Dish', *Du Pont Magazine*, April/May 1952, pp. 1–4

39 Corbman, B. P., *Textiles: Fiber to Fabric*, McGraw-Hill Book Company, New York, 1975, 5th edition, p. 3

40 *The Origin and Development of Courtaulds Between 1700 and 1986*, no author given, Courtaulds publication, p. 67

4
Paris Couture Embraces Man-Mades

1 Morais, Richard, *Pierre Cardin, The Man Who Became a Label*, Bantam Press, London, 1991

2 ibid.

3 Kester, Gordon H., 'Idea Fabrics', *Du Pont Magazine*, vol. 49 no. 5, October/November 1955, p. 4

4 'What's New', *Du Pont Magazine*, October/November 1959, p. 31

5 ibid.

6 op. cit. Kester, pp. 3–6

7 ibid.

8 ibid. p. 3

9 ibid.

10 Denton, Jane, S., 'Report From Paris', *Du Pont Magazine*, June/July 1954

11 *Better Living*, vol. 12, no. 4, 1958. p. 1

12 *ICI Magazine*, April 1960, illustrated Givenchy's glamorous cocktail dress, 'like many other Paris models this season, [it] is made in a "Terylene"'.

13 op. cit. Morais, p. 111

14 ibid.

15 . ibid. p. 112

16 Mehrmann, L., 'Introduction of Qiana through Parisian Couture', submitted to Du Pont on 8 August 1968, pp. 1–20

17 Bennett-England, Rodney, *Dress Optional*, Peter Owen Ltd, London, 1967, p. 94

18 op. cit. Mehrmann, p. 14

19 ibid. pp. 16 & 17

20 ibid. p. 11

21 McGlew, Ray, '40 Million Home Couturiers', *Du Pont Magazine*, January/February 1966, pp. 16–20

22 ibid. p. 18

23 ibid.

5
The Fabric of Pop

1 Barthes, R., *Mythologies*, 1957, reprinted Vintage, London, 1972, p. 97

2 ibid.

3 Katz, Sylvia, *Classic Plastics*, Thames and Hudson, London, 1984

4 See the work of the photographer Roger Mayne, 1956, Southam St Album

5 Abrams, Mark, *Teenage Consumer Spending in 1959*, London Press Exchange, 1961

6 ibid.

7 'Teen Age Boom Rocks the Market', *Du Pont Magazine*, vol. 59, no.5, September/October 1965, p. 4

8 ibid.

9 'The Well-Mannered Look', *Du Pont Magazine*, June 1963, p. 11

10 See Bernard, Barbara, *Fashion in the '60s*, Academy Editions, London, 1978, p. 5

11 Lobenthal, Joel, *Radical Rags, Fashions of the Sixties*, Abbeville Press, New York, 1990, p. 13

12 Ironside, Janey, *Queen*, 21 June 1967, p. 57

13 op. cit. Bernard

14 op cit. Lobenthal, p. 9

15 op cit., Bernard, p. 26

16 Bennett-England, Rodney, *Dress Optional*, Peter Owen Ltd, London, 1967, p. 80

17 ibid.

18 ibid. p. 99

19 During the 1960s, PVC and polythene were the two most widely produced thermoplastics in the world. Although new to fashion, polyvinyl chloride was by no means a new material; it was created in 1835 by a French chemist, Regnault, and was developed and produced in Germany in the 1940s and 1950s.

20 *Quant by Quant*, Cassell and Co., London, 1966, p. 135

21 Ewen, Stuart, *All Consuming Images, The Politics of Style in Contemporary Culture*, Basic Books, 1988, p. 248

22 Fenichell, Stephen, *Plastic, The Making of a Synthetic Century*, Harper Business, HarperCollins, New York, 1996

23 Garland, Madge, *The Changing Form of Fashion*, J. M Dent and Sons Ltd, London, 1970, p. 67

24 op. cit. Ironside, p. 57

6
The Disco Dacron Decade

1 Rivenburg, Roy, 'America's Most Frightening Fabric is Trying to Make a Comeback', *Sunday News Journal*, 6 December 1992, p. 16

2 'Fake Furs', *Du Pont Magazine*, January/February 1970, pp. 17–19

3 op. cit. Rivenburg

4 Schneider, Jane, 'In and Out of Polyester – Desire, Disdain and Global Fibre Competitions', *Anthropology Today*, vol. 10, no. 4, August 1994

5 Seabrook, Jeremy, *The Everlasting Feast*, Allen Lane, London, 1974, p. 237

6 Meikle, Jeffrey, *American Plastic, A Cultural History*, Rutgers University Press, New Jersey, 1995, p. 3

7 Huxley, Aldous, *Brave New World*, Chatto and Windus, London, 1932; author's edition, Penguin Books, London, 1966, p. 39

8 Interview with John Harrison, T.M.S. Partnership, July 1995

9 'Cut these out – in Quant style!', *ICI Magazine*, August 1975, p. 186

10 Barron, Cheryl, 'ICI Fibres and The Crystal Ball of Fashion', *ICI Magazine*, vol. 56, no. 464, November 1978, p. 247

11 ibid. pp. 248–9

12 Siddle, F., 'Terylene Spans The World', *ICI Magazine*, 15 December 1978, pp. 112–22

13 ibid. p. 122

14 Press release held in Du Pont company archives at the Hagley Museum and Library.

15 ibid.

16 ibid.

17 *Du Pont Fashion News*, Du Pont Public Affairs Department, New York, October 1977

18 *Polyester: The Misunderstood Fiber*, press release issued by Du Pont in 1977

19 op. cit. Rivenburg. 'Of the 13 polyester makers that once operated in the United States, only a handful are still in the business. . . . Du Pont is the nation's largest, followed by Hoechst Celanese Corp. (No. 2 in the United States but No. 1 worldwide) and Wellman Inc. These three companies control 90 per cent of the polyester trade in the United States.'

20 *American Fibres and Fabrics*, vol. 3, 1985

21 op. cit. Schneider

22 ibid.

7
Japanese Design and the Fine Art of Technology

1 Rei Kawakubo is in sole control of an international company that trades in 33 countries with a wholesale turnover of $125 million (£80 million) each year. Eleven collections are produced annually, including the signature women's and menswear, eveningwear and shirt lines, furniture and, more recently, fragrance. There are wholly owned boutiques in Tokyo, Paris and New York, as well as a franchise in London and 453 additional points of sale around the world, 390 of which are in Japan. She also finds time to work in other areas, including architecture, interiors, exhibition and graphic design and has produced a series of legendary magazines entitled *Six*.

2 Fax interview with Hiroshi Matsushita, 2 September 1997

3 Cocks, J., 'A Change of Clothes – Designer Issey Miyake Shapes New Forms into Fashion For Tomorrow', *Time*, 27 January 1986, p. 41

4 Oel, L., 'Symbiosis of Handwork and Hightech', interview with Junichi Arai, *kM*, no. 19, Autumn 1996, p. 40

5 ibid. p. 39

6 ibid. p. 38

7 Mimura, Kyoto, 'The Mystical Master of Cloth', *Mainichi Daily News*, 27 February 1989

8 ibid.

9 Paper describing the company history and policy received from Nuno in August 1994.

10 During the early 1990s, Liberty in London stocked the Nuno range of fabrics which sold for around £100 ($150) per metre.

11 Louie, Elaine, 'New Era in Design Materials', *New York Times*, 19 August 1993, p. 9

12 Arad, R. (ed.), *The International Design Yearbook*, 1996, p. 158

13 Arai, Junichi, 'Soft, Metallic Melt-Off', *Look Japan*, October 1997, p. 24

14 The Toyo Rayon Company was founded in 1926. Today, Toray Industries Inc. produces fibres and textiles, plastics, chemicals, advanced composite materials, pharmaceuticals and multimedia, electronic, civil engineering and fashion products.

15 Hongu, T. & Phillips, G. O., *New Fibres*, Ellis Horwood Ltd, Chichester, 1993, p. 6. This publication gives a thorough description of the emergence of the new high-technology fibres in Japan.

16 'Fixing Odours to Textiles', *High Performance Textiles*, Elsevier Science Ltd, August 1994, pp. 7–8

17 op. cit. Hongu & Phillips, p. 66

18 ibid. p. 73

19 op. cit. 'Fixing Odours to Textiles', *High Performance Textiles*

20 On average, a fragrance that retails for £35 ($50) has cost £2 ($3) to produce.

8
Chic Synthetic

1 Rivenburg, Roy, 'America's Most Frightening Fabric is Trying to Make a Comeback', *Sunday News Journal*, 6 December 1992, p. 16

2 *Du Pont Magazine*, vol. 54 no. 2, March/April 1960, pp. 2–4

3 'The Bottom Line in Fashion', *Du Pont Magazine*, November/December 1975, vol. 69, no. 5, p. 1

4 Interview with Georgina Godley, 14 September 1993

5 By 1992 there was sufficient academic material to publish *The Madonna Connection: Representational Politics, Subcultural Identities and Cultural Theory*, ed., Cathy Switchenberg, 1992

6 Steele, Valerie, *Women of Fashion*, Rizzoli, New York, 1991, p. 104

7 ibid. p. 108

8 ibid. p. 114

9 ibid. p. 190

10 ibid.

11 op cit. Rivenburg

12 ibid.

13 ibid.

14 Du Pont granted ICI a licence to produce nylon soon after its invention and has now taken over this ICI division so that it is producing nylon in Europe for the first time. 'Tactel' is a range of nylon yarns which can be transformed into an infinite variety of fabric effects – 'texturals', 'micro', 'diabolo', 'aquator' and 'multisoft' – to make anything from lingerie to fleece jackets. It was first used in action sportswear and for skiwear. The speed with which the word 'microfibre' has entered the mainstream fashion vocabulary is proof that both the fabrics and the marketing campaigns have succeeded in separating the image from the origins, to such an extent that many consumers think that 'micro' is a new generic fibre.

15 Tredre, Roger, *Independent*, 1993

16 Jones, Dylan, *Style*, 23 February 1997, pp. 4–5

17 ibid. p. 5

18 Paradoxically, although synthetic fibres deplete an unrenewable resource, chemically, they are relatively clean when it comes to production. Organic fibres consume many toxic chemicals as both fertilizers and pesticides and during bleaching so that when the 'cradle to the grave' issues are analyzed it is found that natural fibres seriously damage the environment. Cotton accounts for 25 per cent of the world's pesticide consumption and the World Health Organization estimates that cotton causes 1,000,000 cases of acute pesticide poisoning each year and possibly 20,000 deaths.

19 *Du Pont Magazine*, July/August 1971, p. 32. The bikini was 'priced to compete as "disposable" [but] durable enough to last through many swimmings and washings'.

9
Science Fiction Becomes Science Fashion

1 Negroponte, N., *Being Digital*, Hodder and Stoughton, London, 1996, pp. 209–11

2 ibid., in a chapter entitled 'Good Morning, Toaster', pp. 213–15

3 Benedetti, P., & De Hart, N., *On McLuhan*, MIT Press, Cambridge, Massachusetts, 1996, p. 186

4 Ward, M., 'Teardrop of DNA makes Logical Leap', *New Scientist*, no. 2068, 8 February 1997, p. 14

5 Myles Harris, 'Pill-sized Computers: a Hard Act to Swallow', *Evening Standard*, 28 August 1996, p. 28

6 Uhlig, R., 'The Latest Computers You Can Wear Out', *Daily Telegraph*, 23 August 1996

7 Szaniwaski, K., 'Survey of UK Telecommunications Market', *Financial Times*, 21 March 1996

8 Gershenfeld., N., 'Wearable Computing', *Wired*, December 1995, p. 112

9 ibid.

10 Zimmerman, Tom, *Wired*, December 1995

11 Hemblade, Christopher, 'Amazing Technological Dream Clothes', *Guardian*, 23 March 1994, p. 15

12 Prigg, Mark, 'Computers Vanish as Software is Integrated into Clothes', *Financial Times*, 4 February 1996

Bibliography

ABRAMS, M. A., *Teenage Consumer Spending in 1959: Middle Class Boys and Girls*, London Press Exchange, London, 1961

ADANUR, Sabil, *Wellington Sears Handbook of Industrial Textiles*, 1995

ARAI, Junichi, *Hand and Technology*, Arai Creation System, Japan, 1994

ARNOLD, H., *The Story of Rayon*, United Trade Press, 1944

ASH, J. & WILSON, E. (eds), *Chic Thrills: A Fashion Reader*, Pandora, London, 1992

ASH, J. & WRIGHT, L. (eds), *Components of Fashion*, Routledge, London, 1988

BARNARD, M., *Fashion as Communication*, Routledge, London, 1996

BARTHES, R, *Mythologies*, 1957, reprinted Vintage, London, 1972

BARTHES, R., *The Fashion System*, Hill and Wang, New York, 1984

BEERY, P. G., *Stuff: The Story of Materials in the Service of Man*, D. Appleton, New York, 1930

BENDURE, Z. & PFEIFFER, G., *America's Fabrics, Origin and History, Manufacture, Characteristics and Uses*, Macmillan Company, New York, 1946

BENEDETTI, P., & De Hart, N., *On McLuhan*, MIT Press, Cambridge, Massachusetts, 1996

BENNETT, E., J. & SPIEGEL, J. (eds), *Human Factors in Technology*, McGraw Hill, New York, 1963

BENNETT-ENGLAND, Rodney, *Dress Optional*, Peter Owen, London, 1967

BERNARD, Barbara, *Fashion in the '60s*, Academy Editions, London, 1978

BIJKER, E., HUGHES, & PINCH, T. P. (eds), *The Social Construction of Technological Systems: New Directions in the Sociology and History of Technology*, Cambridge, MIT Press, London, 1987

BOCOCK, R, et al., *Consumption and Lifestyles*, Polity, Cambridge, 1992

BOHDANOWICZ, J. & CLAMP, L., *Fashion Marketing*, Routledge, London, 1994

BORDIEU, P. *Distinction*, Harvard University Press, Cambridge, Massachusetts, 1984

BREWER, J. & PORTER, R., *Consumption and the World of Goods*, Routledge, London, 1993

BRISCOE, L., *The Textile and Clothing Industries in the U.K.*, Manchester University Press, 1971

BRITISH NYLON SPINNERS, *Twenty-Five Years of British Nylon Spinners*, 1941–1965 B.N.S., Pontypool, 1965

BRUNNSCHWEILER, D. & HEARLE (eds), *Polyester, 50 Years of Achievement*, The Textile Institute, Manchester, 1993

CARR, H. & POMEROY, J., *Fashion Design and Product Development*, Blackwell, Oxford, 1992

COLEMAN, D.C., *Courtaulds: An Economic and Social History*, vols. 1 & 2, Clarendon Press, London, 1969

COOK, J. Gordon, *Handbook of Textile Fibres, Man-Made Fibres*, Merrow Publishing Co. Ltd, Watford, 1968 (first published 1959)

CORBMAN, B.P., *Textiles: Fiber to Fabric*, McGraw-Hill Book Co., New York, 1975, 5th edition

CRAIK, J., *The Face of Fashion: Cultural Studies in Fashion*, Routledge, London, 1994

CROCKETT, S.R., *An Introduction to Man-Made Fibres*, Sir Isaac Pitman and Sons Ltd, London, 1966

CROSS, G., *Time and Money: the Making of Consumer Culture*, Routledge, London, 1993

DAVIS, F., *Fashion, Culture and Identity*, University of Chicago Press, Chicago, 1992

DAVIS, John, *Youth and the Condition of Britain, Images of Adolescent Conflict*, Athlone Press, 1990

DEITCH, J. & FRIEDMAN, D. (eds), *Artificial Nature*, Deste Foundation for Contemporary Art, Athens, Geneva, New York, 1990

De PAOLA, H. & MUELLER, C., *Marketing Today's Fashion*, Prentice Hall, New Jersey, 1986

DINOTTO, A., *Plastics: Designed for Living*, Abbeville Press, 1984

DOREE, Charles, *The Methods of Cellulose Chemistry*, Van Nostrand, New York, 1933

DOUGLAS, M. & ISHERWOOD, B., *The World of Goods; Towards an Anthropology of Consumption*, Allen Lane, London, 1979

Du Pont, The Autobiography of an American Enterprise, no author given, E. I. Du Pont De Nemours & Co. (Inc.), Wilmington, Delaware, 1952

EASEY, M., *Fashion Marketing*, Blackwell, Oxford, 1994

EPSTEIN, S., & WILLIAMS, B., *Miracles from Microbes*, New Brunswick, New Jersey, Rutgers University Press, 1946

EVANS, C. & THORNTON, M., *Women and Fashion: A New Look*, Quartet Books, London, 1989

EWEN, Stuart, *All Consuming Images, The Politics of Style in Contemporary Culture*, Basic

Books, 1988

FEATHERSTONE, M., *Consumer Culture and Postmodernism*, Sage, London, 1991

FELTWELL, J., *The Story of Silk*, Alan Sutton, Gloucestershire, 1990

FENICHELL, S., *Plastic, The Making of a Synthetic Century*, Harper Collins, New York, 1996

FRASER, G.L., *Textiles By Britain*, George Allen and Unwin Ltd, London, 1948

FRASER, Kennedy, *The Fashionable Mind: Reflections on Fashion*, Alfred A. Knopf, New York, 1981

FRIEDEL, Robert, *Zipper: an Exploration in Novelty*, W.N. Norton, New York, 1995

GALBRAITH, J.K., *The Affluent Society*, Penguin, London, 1962 (first published 1958)

GARLAND, Madge, *The Changing Form of Fashion*, J. M. Dent and Sons Ltd, London, 1970

GARRETT, A.E., *Fibres for Fabrics*, Hodder and Stoughton, London

GREENHALGH, P., *The World Market for Silk*, Tropical Development and Research Institute, London, 1986

GREENWOOD, K. & MURPHEY, M., *Fashion Innovation and Marketing*, Macmillan, London, 1978

HAGUE, D.C., *The Economics of Man-Made Fibres*, Gerald Duckworth, London, 1957

HALL, S. & JEFFERSON, T (eds), *Resistance through Rituals: Youth Subcultures in Post-War Britain*, Hutchinson, London, 1976

HARD, Arnold, *The Story of Rayon*, United Trade Press Ltd, London, 1939

HARRAP, J., *The Rayon Industry in the Inter-War Period, Textile History*, vol. 1, David and Charles, London, 1978

HATCH, K., *Textile Science*, West Publishing Company, Minneapolis, 1993

HAYNES, William, *American Chemical Industry: Decade of New Products*, D. Van Nostrand, New York, 1954

HERMES, Matthew E., *Enough for One Lifetime* (biography of Wallace Carothers), ACS/CHF Books, Du Pont, 1996

HONGU, T. & PHILLIPS, G.O., *New Fibres*, Ellis Horwood Ltd, Chichester, 1993

HOOPER, L., *Silk, It's Production and Manufacture*, Pitman and Sons Ltd

HOUNSHELL, D., & SMITH, J., *Science and Corporate Strategy: Du Pont, R&D, 1902–1980*, Cambridge University Press, Cambridge, 1988

HUXLEY, A., *Brave New World*, Chatto and Windus, London, 1932

I.C.I. Ltd, *Landmarks of the Plastics Industry* (to mark the centenary of Alexander Parkes' invention of the first man-made plastic), Birmingham, 1962

IRONSIDE, J., *A Fashion Alphabet*, Michael Joseph, London, 1968

IRONSIDE, J., *Janey, an Autobiography*, Michael Joseph, London, 1973

JERNIGAN, M. & EASTERLING, C., *Fashion Merchandising and Marketing*, Macmillan, New York, 1990

KATZ, Sylvia, *Classic Plastics*, Thames and Hudson, London, 1984

KOZLOSKI, L., *U.S. Space Gear Outfitting the Astronaut*, Airlife and Smithsonian Institute, 1994

L'ARBOUSSET, M. Laurent de, *On Silk and the Silkworm*, Translated by E. Wardle, Leek, Eaton, 1905

LEWIN, M. & LOBENTHAL, Joel, *Radical Rags, Fashions of the Sixties*, Abbeville Press, New York, 1990

LURY, C., *Consumer Culture*, Polity, Oxford, 1996

McCRACKEN, Grant, *Culture and Consumption: New Approaches to the Symbolic Character of Consumer Goods and Activities*, Indiana University Press, Bloomington, 1988

McLUHAN, Marshall, *The Medium is the Massage*, Penguin Books, London, 1967

McROBBIE, A., *Feminism and Youth Culture: From 'Jackie' to 'Just Seventeen'*, Macmillan, London, 1991

MANZINI, Ezio, *The Materials of Invention: Materials and Design*, Milan, 1986

MEIKLE, Jeffrey, *American Plastic, A Cultural History*, Rutgers University Press, New Jersey, 1995

MENDES, Valerie, *Pierre Cardin, Past Present and Future*, Dirk Nishen, London, 1990

MEYER-LARSEN, W., *Man Made Fibres*, Rowholt, Hamburg, 1972

MILLER, D., *Material Culture and Mass Consumption*, Blackwell, Oxford, 1987

MOFFAT, Peggy & CLAXTON, William, *The Rudi Gernreich Book*, Rizzoli, New York, 1991

MORAIS, R., *Pierre Cardin, The Man Who Became a Label*, Bantam Press, London, 1991

NEGROPONTE, N., *Being Digital*, Hodder and Stoughton, London, 1996

PRESTON. K. (eds), *High Technology Fibres: Handbook of Fibre Science and Technology*, vol. III, Marcel Dekker, New York, 1985

QUANT, M., *Quant by Quant*, Cassell and Co., London, 1966

READER, W. J., *Imperial Chemical Industries: A History*, Oxford University Press, Oxford, 1976

SASSO, John, & BROWN, Michael A., *Plastics in Practice*, McGraw-Hill, New York, 1945

SCHNEIDER, J. & WEINER, A., *The Cloth of Human Experience*, Smithsonian Press, 1989

SCHOTZ, S.P., *Synthetic Rubber*, Van Nostrand, New York, 1927

SEABROOK, Jeremy, *The Everlasting Feast*, Allen Lane, London, 1974

Silk and Rayon Users Association, *The Silk Book*, W. S. Cornell Ltd, 1951

SPARKE, P., *The Plastics Age, From Modernity to Post Modernity*, Victoria and Albert Museum, London, 1990

SPARKE, P., *As Long As It's Pink, The Sexual Politics of Taste*, Pandora, London, 1995

SPITZ, Peter, H., *Petrochemicals: The Rise of an Industry*, John Wiley, New York, 1988

STEELE, Valerie, *Women of Fashion*, Rizzoli, New York, 1991

The History of E.I. du Pont de Nemours Powder Company, no author given, originally published by Business America, New York, 1912, this edition reprinted by Lindsay Publications Inc., Manteno, 1990

The Origin and Development of Courtaulds Between 1700 and 1986, no author given, Courtaulds publication

TISDELL, C.A., & McDONALD, P.W., *Economics of Fibre Markets: Interdependence Between Man-Made Fibres, Wool and Cotton*, Pergamon Press, Oxford, 1979

TOMLINSON, Alan, *Consumption, Identity and Style*, Routledge, London, 1990

WARD-JACKSON, C.H., *A History of Courtaulds*, The Curwen Press (for private circulation)

WATSON, Jacky, *Textiles and the Environment*, The Economist Intelligence Unit, London, 1992

WHITE, Palmer, *Elsa Schiaparelli, Empress of Paris Fashion*, Aurum Press, London, 1986

WILLIAMS, Jon M. & MUIR, Daniel T., *Corporate Images: Photography and the Du Pont Company 1865–1972*, The Hagley Museum And Library, Wilmington, Delaware

WILLIAMSON, Judith, *Consuming Passions: The Dynamics of Popular Culture*, Marion Boyars, London, 1986

Acknowledgements

Over the several years that I have been researching this subject, I have been given help by many very generous people and it is possible here to name only a few. Working with Professor Penny Sparke on the Victoria and Albert Museum's 1989 'Plastics' exhibition first inspired my interest in polymers and stirred my curiosity about the near-forgotten textiles of my childhood. For making this book a reality, I am especially grateful to my editor, Sarah Polden, who possesses superhuman powers of patience, dedication and a fine sense of humour. I must also thank Bloomsbury for taking on such an unusual topic in the first place, Satpaul Bhamra for designing the book and for managing to include so much material, my agent, Rosemary Canter, for her support and advice, Laura Fidment for help with picture research and Peter Barber for producing such a thorough and useful index.

I am deeply indebted to the Design Council in London for its funding which launched the research in 1994, and in particular to Richard Shearman and Jeremy Myerson for giving the original proposal their support. Other related awards that were much appreciated came from the Apparel Trust at the Textile Institute and from the Hagley Center in Wilmington, Delaware.

At the Royal College of Art I have been fortunate to work with exceptionally inspiring colleagues and students within the School of Fashion and Textiles and I am particularly grateful to Professor John Miles and to Professor James Park, for their unlimited professional encouragement and valued friendship. I owe a very special thank-you to several friends and (mostly) former students: Carolyn Corben, Harvey Bertram-Brown, Deborah Milner, Jo Gordon, Zowie Broach, Brian Kirkby, Paul Helbers, Dimitry Krul, Maya Arazi, John Crummay, Janet Stoyal, Lorna Ross and to Justin Oh for bringing Sarah Polden and myself together. The RCA Library staff have provided invaluable help and I must thank Jan Murton, Ann George, Pauline Rae, Eugene Rae, Lucy Neville, Nichola Marry and Robin Cousins. From the very beginning, I have been kept going by the friendly encouragement of Suzy Menkes, Fashion Director at the *International Herald Tribune*, David Shah of Metropolitan Publishing, and Stephen Higginson of *International Textiles*.

The major textile-fibre producers, Courtaulds, Du Pont and ICI, have been of enormous assistance and I would like to thank David Wilkinson, Carol Felton, David Service, Simon Smith, Caroline Powell, Maurice Unwin, Julia Burrows and Joanna Bowring at Courtaulds plc and Sheena Russell, archivist at ICI. Du Pont is an impressively vast organization and many people have replied to all sorts of inquiries, including Robin Noakes, Global Brand Manager, Tactel, Robert Pruyn, Business Director, North America, and Robin Harrington of the Public Relations Department, E.I. Du Pont de Nemours & Co., Du Pont, Wilmington.

Karen Jones, Marketing Manager, Du Pont UK, and Lisa Waughman of Caroline Neville Associates have also supplied me with very useful material.

Extensive Du Pont archives are held at the Hagley Museum and Library in Wilmington and it was the opportunity to carry our research there that really clarified and illuminated the human story of synthetic fibres for me. Access to this material was made astonishingly easy, thanks to the accommodating and considerate staff at the Center, and it proved to be a watershed for the book. I would especially like to thank Dr Roger Horowitz, Michael Nash, Susan Hengel, Lynn Joshi, Deborah Hughes, Carol Lockman and (as the number of picture credits demonstrates) I owe my greatest debt to Jon Williams and to Barbara Hall of the pictorial department.

My sincere thanks go to Dr Pamela Golbin, Curator of Costume at the Musée des Arts de la Mode et du Textile, Paris, who has been a very generous source of unique information and perceptive comment and who gave me such liberal access to the collections and archives of the museum. For their kind individual contributions I am also grateful to Mikel Rosen, Franca Sozzani, Director of Italian *Vogue*, Joanna Marschner of the Court Dress Collection, Kensington Palace, Akiko Fukai, Director, Kyoto Costume Institute, Rosemary Harden of the Bath Museum of Costume, Dr Susan Mossman of the Science Museum, London, Professor Neil Gershenfeld, Director, Massachusetts Institute of Technology, Professor Jeffrey Meikle of the University of Texas, Peter Cochrane, Head of Research, B. T. Laboratories, John Harrison and Roy Hollis of TMS Partnership, Colin M. Purvis of the European Man-Made Fibres Association, Elsbeth Juda, Ann Sutton, Rachel Worth, Jill Lawrence, Denise Ford, Peter King, Susan Breakell, Company Archivist, and Charlotte Smith at Marks and Spencer, Mr T. Hosokawa of Toray Textiles, Mr T. Fujimoto of Unitika, Reiko Sudo of Nuno Textiles, Junichi Arai, Annie Quesnel at Jakob Schlaepfer, Jelka Music at Comme des Garçons, Mr Matushita, Laurent Cassagnau at Girault Taboo, Theresa Saibene of Ratti Silks, Ann McGovern and Céline Clot-Pollisse at Jean Paul Gaultier, Isobel de Lazan Derya at Pierre Cardin, Rei Kawakubo, Richard Nott, Graham Fraser, Lisa Bruce, Lainey Keogh, Mary Quant, Georgina Godley, Nigel Atkinson, Helen Storey, Jo Hunter and Adam Thorpe of Vexed Generation, and Emma Wheeler at Nick Knight.

Very special thanks go to all the designers and photographers – most conspicuously Chris Moore – whose work is illustrated here, breathing the life of fashion into the fibre story. I am particularly happy to express my gratitude to Steve Hiett and Nick Knight for the privilege of including their work.

Finally, all my love and heartfelt thanks go to my husband, Roger, whose photographic talents in Delaware and elsewhere gave me almost half the images reproduced in the book and whose patience and good humour have seen him through what must have seemed an eternity of fibrespeak and fashion chatter.

Picture Credits

The publishers have made every effort to include full acknowledgements for the illustrations but any further information would be gratefully received.

The illustrations are listed by source or copyright holder and by page number. Where there is more than one illustration on a page, the following abbreviations have been used: l: left, r: right, c: centre, t: top, b: bottom. Where all the illustrations on a page or in a row come from the same source, this has been indicated by (all).

Advertising Archive 55t 69b 110bl 119b 127
The Ambassador magazine 57l 75
American Fabrics magazine 66b
Junichi Arai 137t (all)
Maya Arazi 168 181
Nigel Atkinson 146 166bl
Photographer Christian Behnke 167bl
Photographer Piero Biasion 2 156c (all)
Photograph by Blumenfeld. Courtesy Vogue 1953 © copyright 1953 (renewed 1981) by the Conde Nast Publications, Inc. 67
Liza Bruce/Dan Lecca 161b
Photograph by Anton Bruehi. Courtesy Vogue 1939 © copyright 1939 (renewed 1967, 1995) by the Conde Nast Publications, Inc. 178
Canal + Image UK Ltd 56
Hussein Chalayan 166br
Comme des Garçons 131 131c
Courtaulds 21b 22
William R. Culver © National Geographic Society 43
Photographer Louise Dahl-Wolfe 160
Marion Foale 104t
Fortune magazine 29
Photographer Gry Garness 167br
Jean Paul Gaultier 158t 159tl 167tc
Getty Images 78t 118t 126
Courtesy of the Hagley Museum and Library 13 14 15 24 25t 26 30 32 33 35 36 37 38 39 41 42t 42b 45 46 47 49 (all) 50 51 52 57c 57r 59 60 61 62 63 64 65 68 69t 71 76 79t 79b 80 81 (all) 82t 82b 83 84 (all) 85 86 (all) 91tl 94t 94b 97t 97b 103t 110br 119t 121 123 125 (all)
Susannah Handley 21t 105
Harper's Bazaar 78b
Photographer Steve Hiett 144
ICI Fibres Ltd 58b 73 74 101t 101b 112tl 112r 122 124
ICI Fibres Ltd/Royal College of Art 98 115
Inch 167tr
Photographer Suresh Karadia 167cl 167c
Lainey Keogh 167tl
Photographer Nick Knight 128 130 154–5
Life magazine 58t
Reproduced by kind permission of Macclesfield Museums Trust 16b
Guy Marineau/Java 139 (all) 143 175t
Marks & Spencer Company Archive 70 147
Metropolitan Museum of Art, New York 10
Musée des Arts de la Mode et du Textile, Paris, collection UFAC 9 91r, t–b 91inset, bl
Photographer Chris Moore 6 132t 132b 133 134 135 148 149tl 149bl 149br 150 151 (all) 156tl 156tr 156b (all) 157 159tc 159c 159cr 159bc 161t

162 (all) 166t 172 173 175c 175b 176 177
Courtesy of Museo Tessile, Fondazione Antonio Ratti 16t (group)
The New RenaisCAnce 152, 153 156tc
Nova magazine 90 109b
Paramount Pictures courtesy of the Ronald Grant Archive 116
Picture Post 44
Popperfoto 92
Mary Quant 106 108
Zandra Rhodes 113
John Richmond 159br
Royal College of Art 72 114 171
Photographer Franck Sauvaire 174
Jakob Schlaepfer 158b 159cl
Science Museum, London/Science and Society Picture Library 12 19 96
Shoe Museum, Toronto Foundation 89br
Helen Storey 149tr 159bl 170
Reiko Sudo 137c (all) 137b (all)
Photographer Yoshi Takata 89t 89bl
Tencel 163 164
Town magazine 112bl
Tristan Webber 159tr
Photographer Stuart Weston 165 167cr
Wolverhampton Art Gallery, West Midlands, UK/Bridgeman Art Library, London/New York 17

Index

Page numbers in *italic* refer to captions to illustrations. Notes are indicated by a page number with *n* (eg 183n)